A SHORT GUIDE to the Human Genome

Related Titles from Cold Spring Harbor Laboratory Press

Bioinformatics: Sequence and Genome Analysis, 2nd Edition
The Dog and Its Genome (Cold Spring Harbor Monograph Series 44)
Genetic Variation: A Laboratory Manual
The Genome of *Homo sapiens* (Cold Spring Harbor Symposia on Quantitative Biology LXVIII)
Genomes (Cold Spring Harbor Monograph Series 46)
Molecular Biology of the Gene, 6th Edition
Recombinant DNA: Genes and Genomes–A Short Course, 3rd Edition
Won for All: How the *Drosophila* Genome Was Sequenced

Handbooks

At the Bench: A Laboratory Navigator, Updated Edition
At the Helm: A Laboratory Navigator
Binding and Kinetics for Molecular Biologists
Experimental Design for Biologists
Lab Math: A Handbook of Measurements, Calculations, and Other Quantitative Skills for Use at the Bench
Lab Ref, Volume 1: A Handbook of Recipes, Reagents, and Other Reference Tools for Use at the Bench
Lab Ref, Volume 2: A Handbook of Recipes, Reagents, and Other Reference Tools for Use at the Bench

A SHORT GUIDE to the Human Genome

Stewart Scherer

COLD SPRING HARBOR LABORATORY PRESS
Cold Spring Harbor, New York • www.cshlpress.com

A Short Guide to the Human Genome

Published by Cold Spring Harbor Laboratory Press
Printed in the United States of America

Publisher	John Inglis
Acquisition Editor	Alexander Gann
Development, Marketing, & Sales Director	Jan Argentine
Developmental Editor	Maria Smit
Project Coordinator	Mary Cozza
Production Editor	Kaaren Hegquist
Desktop Editor	Lauren Heller
Production Manager	Denise Weiss
Book Marketing Manager	Ingrid Benirschke
Sales Account Manager	Elizabeth Powers
Cover Designer	Ed Atkeson

Front cover artwork: By permission. From *Merriam-Webster's Collegiate® Dictionary, 11th Edition* © 2007 by *Merriam-Webster, Incorporated (www.Merriam-Webster.com).*

Library of Congress Cataloging-in-Publication Data

Scherer, Stewart.
A short guide to the human genome / Stewart Scherer.
p. cm.
Includes bibliographical references and index.
ISBN 978-0-87969-791-4 (pbk. : alk. paper)
1. Human genome. I. Title.

QH431.S34 2008
599.93'5--dc22

2008005705

10 9 8 7 6 5 4 3 2 1

All Cold Spring Harbor Laboratory Press publications may be ordered directly from Cold Spring Harbor Laboratory Press, 500 Sunnyside Blvd., Woodbury, New York 11797-2924. Phone: 1-800-843-4388 in Continental U.S. and Canada. All other locations: (516) 422-4100. FAX: (516) 422-4097. E-mail: cshpress@cshl.edu. For a complete catalog of all Cold Spring Harbor Laboratory Press publications, visit our World Wide Web Site http://www.cshlpress.com/.

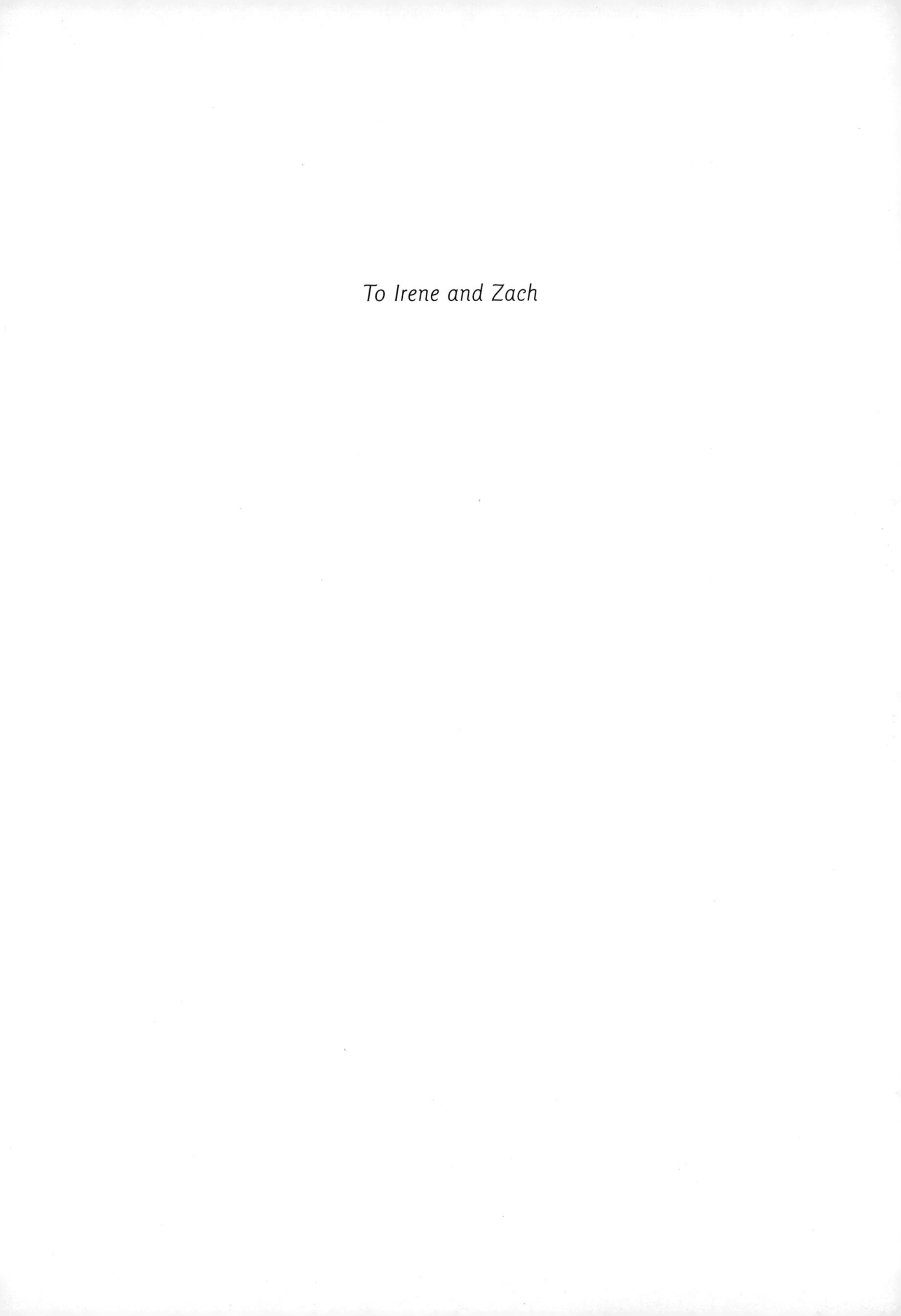

To Irene and Zach

CONTENTS

DETAILED CONTENTS

PREFACE

THE COMPLETION OF THE HUMAN GENOME SEQUENCE is one of the great accomplishments of the biological sciences. Yet, even with the complete sequence and its extensive annotation, it can be surprisingly difficult to find answers to seemingly simple questions about the genome for a seminar, manuscript, or classroom lecture. Often, one is certain the information to answer the question exists, but perhaps the information is not organized or reduced to a useful or accessible form.

While working on a more comprehensive text about the human genome, I had to assemble information related to many of these commonly asked questions. Generally, each question has what one might describe as the short answer and the long answer, the latter including the many caveats and explanations about data sources and methods of analysis. In this small volume, I try to provide a bit of both. In some cases, the answers come from calculations starting with the complete genome. In others, I feel that specific examples may be more useful than comprehensive statistics. Some of the sections revisit familiar subjects; others present some aspects of human biology that receive less attention. Where possible, I have included graphical presentations to complement more detailed information in the tables.

This work covers many areas of human biology. Rather than comprehensively reviewing the literature on all of these topics, the intent has been to show how one can start with relatively little information other than the annotated genome sequence and build towards more general conclusions. Even if you do not find the answer to your question in this volume, I hope you find some interesting biology during your search.

Please send feedback on this book. To keep this a *Short Guide*, coverage of some topics in genomics and other areas is relatively limited. Feel free to suggest additional topics for consideration in future editions. A Web site at http://www.cshlpress.com/genome has been set up for this purpose.

I would like to thank Alex Gann and Cold Spring Harbor Laboratory Press for suggesting that a small volume such as this might be a useful addition to the works available for understanding the human genome. Jasper Rine had many helpful ideas for topics to include in the volume. Special thanks go to David Lipman, Kim Pruitt, Deanna Church, Donna Maglott, and Tom Madden at NCBI for their assistance over the years in helping me to make optimal use of the tools and databases that are available through NCBI. Michael Brent, Mark Johnston, Eugene Koonin, and Igor Rogozin helped answer questions that arose while collecting the information in some of the tables. John Inglis, Maria Smit, Mary Cozza, Kaaren Hegquist, Lauren Heller, Denise Weiss, and the staff at CSHL Press turned the idea of a book like this into a reality. Finally, I would like to thank my family for their support with the project from beginning to end.

STEW SCHERER
February, 2008

CHAPTER ONE

INTRODUCTION

SINCE THE ANNOUNCEMENT OF THE DRAFT HUMAN GENOME SEQUENCES in 2001, much progress has been made toward a complete sequence and comprehensive annotation of the genome. Aside from the repeated regions where sequences are not likely to be detailed, relatively few regions have significant ambiguities. The annotation for protein-coding regions is extensive, and the annotation for transcripts that do not encode proteins is growing. This information is widely available through the public databases and their associated map-viewing tools. However, even with these resources, it is not always easy to find the answers to seemingly straightforward questions about the genome.

Within each of the following chapters, there are a series of questions and answers that cover broad subject areas in genomics. Most of the answers include tabular or graphical representations of results derived from the genomic information. The accompanying text and explanatory notes give a flavor for the many choices of data sets and methods that are needed to present a concise answer to the question posed.

The answers to many questions require specialized calculations starting from raw sequences, database tables, and related information that underlie familiar tools. A large fraction of the tables and figures in this work were generated in this fashion.

A second class of answers requires manual assembly of relevant data. Often, these can be assisted by computer searches for keywords and sequence similarity. After the data of interest are collected, additional computation may be needed to render the information in a more useful form. Many of the figures in this volume were developed in this way.

Other answers require extensive manual assembly of information on specialized topics that rely heavily on experimental (rather than computational) analyses. The sections of this work that answer this class of question draw most of their information from the literature.

In assembling this volume, one of the first steps was the selection of comprehensive data sources on the human genome. Two draft human genome sequences were published in 2001. Although these sequences were in general agreement, reconciling them at a detailed level is not trivial. They may differ because of sequencing errors, assembly issues, or polymorphisms. Additional sequence data exists for portions of the genome, and the amount of raw human sequence information has continued to grow. For this volume, the starting point has been reference assembly 36.2 of the human genome from the National Center for Biotechnology Information (NCBI). In this assembly, which was released in 2006, very little sequence data is not located on the 25 sequences (22 autosomes, X chromosome, Y chromosome, and mitochondrial genome). Most of the ambiguity resides in large blocks of repeated sequences.

Having a reference set of sequences in hand is only part of what is required to answer most questions about the genome. For information about genes, again, there are multiple sources of information. A very large amount of information is annotated onto the genome for use in map-viewing tools. This includes tables of transcripts with their exon–intron structure, repeated sequences, and other features. In addition, there are large numbers of computational predictions of human genes. Although many of these predictions may prove to be pseudogenes, others represent the human homologs of known genes in other species. Still others may prove to have novel functions.

Attaching transcripts to the reference genome sequence is only one step in connecting them to their associated biology. NCBI maintains a second set of transcripts and protein products in the form of its RefSeq database. These are manually curated entries that are linked to the literature and other data sources. Although these two sets of information are largely congruent, some differences remain, often in parts of the genome where questions remain about the DNA sequence. Many of the tables and figures in the text were generated from the genome sequence and map, but in some cases the RefSeq collection that was available concurrently with annotation 36.2 of the genome sequence in September of 2006 was used.

Over time, the varied data sources will converge on a relatively stable, annotated, reference genome. Although the data available at this writing are not yet there, they are quite close. As our knowledge about the genome continues to grow, the information presented here should change mainly in detail.

The figures in this volume were prepared starting with the coordinates from the genome annotation or with results from sequence comparisons. Although many of the figures are relatively straightforward maps, histograms, and scatter plots, a few of the methods involved in generating the figures require some additional explanation.

For some of the figures, such as the figure related to CpG islands on page 37, a conventional histogram was not used; instead, cumulative results were plotted. Such plots are essentially the integral of a histogram where the value on the *y*-axis indicates the percentile rank of the value on the *x*-axis. They are well-suited to certain types of genome data where the distributions have very long tails. It is possible to present multiple plots on a single figure, and one can easily see the fraction of the data that are in a particular size class. These plots are less suited to situations in which the mode or other small peaks are of interest.

In several sections, a graphical presentation is used to show consensus sequences. In these figures, the *x*-axis is simply the coordinate along the sequence. The *y*-axis displays the possible results: the nucleotides or amino acids that might be present. For each position along the *x*-axis, the fractional use of each nucleotide or amino acid is determined and then converted to a box, where black represents 100% use and levels of gray denote fractional levels of use. The levels of gray are linear and continuous to a small baseline level. This baseline makes the rarer choices easier to see against the white background. If the nucleotide or amino acid is not used, there is no box.

Another type of figure is used many times in chapter 10 to describe protein similarity. Rather than using simple percent identity or similarity, a relative BLASTP HSP score is used. A value of 25 is deducted from all results, including self-matches, to adjust for the typical random matches that occur in searches of protein databases of this size, where no significant matches are present (in rare cases, where this adjustment produces small negative scores, they are set to zero). This correction is significant for relatively small proteins. The ratio of the match of interest to the self-match gives a measure of similarity ranging from 0 to 1. This relative score provides weighting for the nature and extent of the match. The method is not well-suited to proteins that do not yield a single aligned region, even with the gapped alignments produced by BLASTP. The scales on the *x*-axes of these graphs do not indicate evolutionary distance; the species of interest are simply evenly spaced.

Finally, the notes at the end of each section provide additional details and information on how the data were collected and presented. Many explain the inclusion or exclusion of specific sequences from the calculations. The references provide sources of information other than the sequence data sets mentioned above and additional context in cases where the short answers provided here form only part of the story.

Data Sources, Methods, and References

The draft genome sequences were presented in:

Lander E.S. et al. 2001. Initial sequencing and analysis of the human genome. *Nature* **409:** 860–921.

Venter J.C. et al. 2001. The sequence of the human genome. *Science* **291:** 1304–1351.

An update on the genome sequence is described in:

International Human Genome Sequencing Consortium. 2004. Finishing the euchromatic sequence of the human genome. *Nature* **431:** 931–945.

For details about the NCBI human genome builds (in particular, Build 36.2), see the release notes that are detailed at http://www.ncbi.nlm.nih.gov/genome/guide/human/release_notes.html.

For a summary of the NCBI databases, including the Map Viewer and RefSeq, see:

The NCBI Handbook [Internet]. Bethesda (MD): National Library of Medicine (US), National Center for Biotechnology Information; 2002 Oct. Available from http://www.ncbi.nlm.nih.gov/books/bv.fcgi?rid=handbook.TOC&depth=2.

BLASTP searches were generally performed without low-complexity filtering. For additional information about BLAST, see:

Altschul S.F. et al. 1997. Gapped BLAST and PSI-BLAST: A new generation of protein database search programs. *Nucleic Acids Res.* **25:** 3389–3402.

For simplicity, specific sequences are referenced by their NCBI GI identifiers in the text. These identifiers can be used to locate the database entry for the exact sequence that was used in the calculation. Those entries also contain the accession numbers and database sources for the sequences.

CHAPTER TWO

DNA AND THE CHROMOSOMES

THIS CHAPTER FOCUSES ON THE BASIC PARAMETERS of the reference human genome sequence, the sizes of the chromosomes, and its nucleotide composition. It also discusses the low-complexity sequences of the genome, the centromeres, and the telomeres.

When the chromosomes are considered individually, one finds significant differences in base composition, even among the autosomes. Later in this volume, some of these compositional differences are further explored through their connection to the distribution of genes on the chromosomes.

Many of the tables in this chapter present values that were computed directly from the 25 reference sequences (the 22 autosomes, the X and Y chromosomes, and the mitochondrial genome). In the reference sequences, there are large blocks of repetitive DNA where sequences have not been fully resolved and are reported as Ns. Therefore, it is important to understand that the calculations were performed on the sequenced genome, rather than on the complete genome.

What Is the Size of Each Chromosome?

Although the human genome sequence is largely complete, the sizes of the chromosomes remain estimates. This is because blocks of repeated sequences have not been fully resolved, and these blocks are counted as unsequenced portions of the genome. The table below presents the estimated sizes of each chromosome, including the sequenced and unsequenced fractions.

Chromosome	Sequenced (Mb)	Unsequenced (Mb)	Total Size (Mb)	Fraction Sequenced (%)
1	225.0	22.2	247.2	91.0
2	237.7	5.2	243.0	97.8
3	194.7	4.8	199.5	97.6
4	187.3	4.0	191.3	97.9
5	177.7	3.2	180.9	98.3
6	167.3	3.6	170.9	97.9
7	155.0	3.9	158.8	97.6
8	142.6	3.7	146.3	97.5
9	120.1	20.1	140.3	85.6
10	131.6	3.8	135.4	97.2
11	131.1	3.3	134.5	97.5
12	130.3	2.0	132.3	98.5
13	95.6	18.6	114.1	83.7
14	88.3	18.1	106.4	83.0
15	81.3	19.0	100.3	81.1
16	78.9	9.9	88.8	88.8
17	77.8	1.0	78.8	98.8
18	74.7	1.5	76.1	98.1
19	55.8	8.0	63.8	87.4
20	59.5	2.9	62.4	95.3
21	34.2	12.8	46.9	72.8
22	34.9	14.8	49.7	70.1
X	151.1	3.9	154.9	97.5
Y	25.7	32.1	57.8	44.4
Total	2858.0	222.4	3080.4	92.8

Although the autosomes are generally numbered according to size (largest to smallest), chromosome 22 is larger than chromosome 21. Both chromosomes 21 and 22 are smaller than the Y chromosome. Chromosome 1 is about 8% of the genome.

For a few of the chromosomes, the sequenced fraction approaches 99%. The Y chromosome has, by far, the largest unsequenced fraction of all chromosomes.

Below is a chromosome map showing unsequenced regions that are 1 kb or longer. For regions 1 Mb or longer, sizes are indicated. The low-numbered ends of the sequences of chromosomes 13, 14, 15, 21, and 22 are the sites of the nucleolus organizers (see p. 46). Many of the other large unsequenced blocks are centromeric heterochromatin.

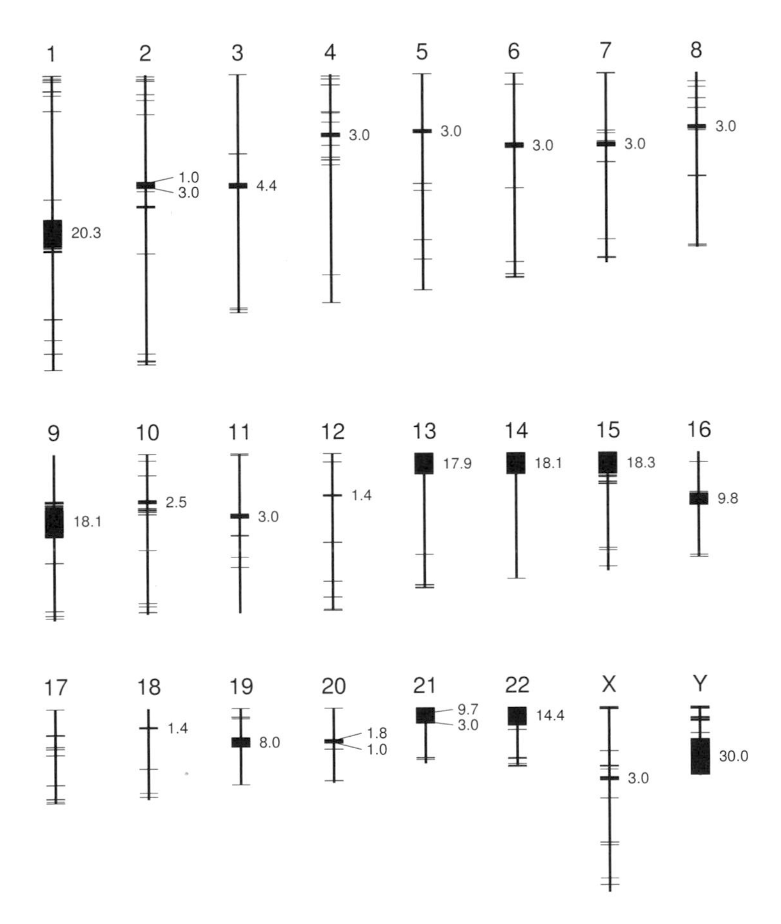

Data Sources, Methods, and References

The data in the table were generated from release 36.2 of the reference human genome sequence. All positions with an A, G, C, or T were counted in the sequenced fraction. In a number of cases, the reported sequences ended with a large block of Ns. Many repeated regions were also large blocks of Ns. Isolated Ns and other ambiguity codes were counted in the unsequenced fraction, whether they were present at a single position or as a large block. A very small fraction of the genomic sequence has not been placed on the chromosomes. It has been omitted from the results.

The figure on the previous page was produced with the same information used to generate the table. The chromosomes and unsequenced regions are drawn to scale, but because of the resolution of the figure, adjacent segments of unsequenced DNA appear as merged in some cases.

What Is the Base Composition of the Nuclear Genome?

The table below presents the proportions of A, G, C, and T in the sequenced fraction of the nuclear genome. The values reported are from the strand presented in the reference genome sequences. Because the unsequenced fraction is small and relatively similar in composition, these numbers reasonably reflect the complete genome.

Nucleotide	Fraction of Sequenced Genome (%)
A	29.53
G	20.46
C	20.44
T	29.57

At the genome level (shown above) and at the chromosome level (not shown), there is no strong strand bias with regard to base composition. This is evident from the table above: The fractions of A and T are nearly equal, and the fractions of G and C are nearly equal.

The mitochondrial genome has a much different base composition than the nuclear genome and shows strong strand bias (see p. 10 for details).

Data Sources, Methods, and References

This table was generated from release 36.2 of the reference genome as described on page 7. Positions with ambiguities were omitted from the counts.

What Is the Base Composition of the Mitochondrial Genome?

The circular reference human mitochondrial DNA sequence is 16,571 nucleotides. The proportions of A, G, C, and T in this sequence are presented below.

Nucleotide	Fraction of Mitochondrial Genome (%)
A	30.86
G	13.16
C	31.33
T	24.66

The mitochondrial genome has a very different base composition than the nuclear genome (see p. 9). It also shows strong strand bias. This is because of the dense packing of the mitochondrial genes and the encoding of almost all mitochondrial genes on the same strand. The gene content and organization of the mitochondrial genome is detailed in chapter 3.

Data Sources, Methods, and References

The values in the table were derived from the mitochondrial sequence of one African (Yoruban) individual (NCBI GI:17981852). See also GI:1153155570 for an alternate reference mitochondrial genome sequence.

What Is the Frequency of Each Dinucleotide?

The frequency of each dinucleotide in the sequenced fraction of the genome is presented in the table below.

Dinucleotide	Fraction (%)
AA	9.78
AG	6.99
AC	5.03
AT	7.73
GA	5.93
GG	5.21
GC	4.27
GT	5.05
CA	7.25
CG	0.99
CC	5.21
CT	7.00
TA	6.57
TG	7.27
TC	5.94
TT	9.80

Most notably, there is a depressed level of the CG dinucleotide.

There is no notable strand bias with respect to dinucleotide frequencies. As shown in the table above, the reverse complements are present at similar levels (e.g., the frequency of AG is very close to that of CT).

Data Sources, Methods, and References

This table was generated from release 36.2 of the reference genome as described on page 7. Positions with ambiguities were omitted from the counts.

Does Base Composition Vary Among the Chromosomes?

The chromosomes have significant differences in their guanine + cytosine (G+C) fraction. These differences are associated with additional variation in the CG dinucleotide percentages for each chromosome.

Chromosome	Nucleotides C+G (%)	Dinucleotide CG (%)
1	41.7	1.01
2	40.2	0.91
3	39.7	0.83
4	38.2	0.78
5	39.5	0.85
6	39.6	0.88
7	40.7	1.01
8	40.2	0.92
9	41.3	1.02
10	41.6	1.03
11	41.6	0.98
12	40.8	0.98
13	38.5	0.84
14	40.9	0.97
15	42.2	1.07
16	44.8	1.39
17	45.5	1.48
18	39.8	0.91
19	48.4	1.89
20	44.1	1.21
21	40.9	1.09
22	48.0	1.66
X	39.5	0.82
Y	40.0	0.85

Differences in G+C are associated with differences in gene density (p. 21).

Data Sources, Methods, and References

The values in the table represent the sequenced fraction of the reference genome sequence (release 36.2) as described on page 7.

What Are the Frequencies and Sizes of Simple Sequence Repeats?

A large number of simple sequence repeats (SSRs) have been annotated on the reference genome. The table below provides some information on the simplest classes of SSRs: homopolymer tracts, dinucleotide repeats, and trinucleotide repeats.

Repeat Type	Count	Mean Size (bp)
A	27,369	28
C	187	32
TA	58,498	85
TG	104,403	51
TC	18,110	39
CG	128	57
ATG	4319	65
CAA	11,002	29
CAG	1799	46
CCA	2447	105
CCG	2275	56
CGA	36	57
CTA	233	63
GAA	1526	78
GGA	2223	53
TAA	16,301	35

There is considerable variation in the frequencies and sizes of these SSRs.

The most abundant classes of trinucleotide repeats (CAA and TAA) are, on average, shorter than the others. The repeats of the trinucleotide CGA are very rare. This is not surprising given the paucity of the CG dinucleotide in the genome (see p. 12). However, repeats of the other class of CG-containing trinucleotides (CCG) are much more common.

Many other more complex repeats are present in the human genome. These include the satellite sequences described in the section on centromeres (p. 15).

Data Sources, Methods, and References

Repeats annotated on one strand of the human genome sequence (release 36.2) were used. Some of the annotated repeats have overlapping coordinates, and these were counted separately. Dinucleotide and trinucleotide repeats of the same base were counted as homopolymers. The reverse complements of each repeat type were merged (e.g., the total for A is the sum of A and T; the total for TG is the sum of the TG and CA repeats). Permuted sequences were also included in the totals (e.g., TG includes GT).

Some relatively short repeats (fewer than 20 nucleotides) are annotated on the chromosomes and are included in the totals reported here. In some cases, they are adjacent to related repeat sequences. The repeats annotated as fewer than 20 nucleotides represent about 0.5% of the homopolymer and dinucleotide repeat counts, except for TA (which comprise about 1%). After removing these repeats, the adjusted totals would be 27,279 for A; 186 for C; 57,858 for TA; 104,067 for TG; 18,013 for TC; and 128 for CG. The mean sizes are minimally affected. These short repeats are an even smaller component for the trinucleotides.

Which Sequences Are Present at the Centromeres?

The α satellite is found at all of the centromeres. Other classes of repeated DNA sequences are also associated with centromeres. The table below lists the sizes of some of the longer classes of repeats, including the α satellite.

Family	Length of Repeat Unit (bp)
α satellite	171
β satellite	68
γ satellite	220

There are many other satellite sequences with complex patterns of shorter repeats. These include the classical satellites described in the references listed below.

Satellite DNAs and related sequences may be found at other locations in the genome. They are often present in large tandem arrays with complex patterns of repeats and variants.

Data Sources, Methods, and References

For examples of the repeats, see GI:36349 (an α satellite consensus), GI:337814 (a β satellite consensus), and GI:1223742 (a γ-type array from the X chromosome).

See also:

Jeanpierre M. 1994. Human satellites 2 and 3. *Ann. Genet.* **37:** 163–171.

Prosser J. et al. 1986. Sequence relationships of three human satellite DNAs. *J. Mol. Biol.* **187:** 145–155.

Rudd M.K. et al. 2006. The evolutionary dynamics of alpha-satellite. *Genome Res.* **16:** 88–96.

What Are the Sequences at the Telomeres?

Human telomeres are usually described as repeats of TTAGGG (or variants of this sequence) on the G-rich strand. However, fewer than one third of the reference chromosome sequences have these canonical repeats (or close variants) at their ends. More than half of the chromosome ends are reported with blocks of Ns, and a few have unrelated sequences.

Telomeres are extended by the action of telomerase, a ribonucleoprotein that contains a protein encoded by *TERT* (telomerase reverse transcriptase) on chromosome 5 and an RNA encoded by *TERC* (telomerase RNA component) on chromosome 3. Sequences closely related to *TERC* are not found elsewhere in the reference genome sequence in searches using BLASTN. In contrast, many small nuclear RNA (snRNA)-related sequences are present in the genome (see p. 58).

Near the 5′ end of the *TERC*-encoded RNA sequence (GI:38176147) is a region complementary to the G-rich telomere strand. This region is diagrammed below (the coordinates refer to GI:38176147).

```
Chromosomal DNA    5′-TTAGGGTTAGGGTTAG-3′
                        |||||||||||
Telomerase RNA     3′-AGAGUCAAUCCCAAUCUGUUU-5′
                          ^         ^
                          56        46
```

After telomere elongation, there is an 11-nucleotide sequence of potential pairing between the DNA and RNA, and the six nucleotides at the 3′ end of the DNA are permuted relative to TTAGGG.

Data Sources, Methods, and References

For the protein sequence of TERT isoform 1, see GI:109633031.

See also:

Gavory G. et al. 2002. Minimum length requirement of the alignment domain of human telomerase RNA to sustain catalytic activity in vitro. *Nucleic Acids Res.* **30:** 4470–4480.

CHAPTER THREE

GENES

THIS CHAPTER FOCUSES ON THE MAJOR QUESTIONS of gene organization and structure. It begins with one of the most commonly asked and difficult to answer questions: What is the number of protein-coding genes in the human genome?

A complementary question related to gene number is the count of pseudogenes. Due to the fact that human genes are generally related to genes in other species (or are in families whose human members have considerable sequence similarity), estimating the total number of genes and pseudogenes is not so difficult. Parsing them into the two categories is the greater challenge.

Genes have a decidedly nonrandom distribution on the chromosomes, and there is enormous variation in gene size and intron–exon structure. After describing some extreme cases, the chapter deals with factors that correlate with gene size and intron number. Also included are discussions of features such as CpG islands, the sizes of untranslated regions (UTRs), and the presence of genes within genes. The chapter concludes with information about the genes encoded by the mitochondrial DNA.

How Many Protein-coding Genes Are Present in the Genome?

This is one of the central and most complex questions about the genome. The following table shows several ways to count human genes.

	NCBI Build 36.2 of the Human Genome Sequence				
	No. of Mapped Genes	– Duplicated Gene Names	+ Unplaced Genes	= Total No. of Genes	Total No. of Genes in RefSeq
All Genes	22,585	27	182	22,740	25,685
All Genes Minus Predicted Transcripts	18,349	22	30	18,357	18,399

This table compares two data sources: One source is the set of genes assigned to the reference chromosome sequences (release 36.2 of the human genome sequence), and the second source is the human RefSeq collection that was available at the same time. When efforts are made to join these data sets, a number of discrepancies remain.

For the purpose of this table, pseudogenes and mitochondrial products (13 protein-coding genes) were excluded from the count of "All Genes" (for both data sets), as were genes not associated with a mapped transcript. Numbers in the "All Genes" row are likely overestimates because some gene fragments are probably pseudogenes. In contrast, the "All Genes Minus Predicted Transcripts" row (which removes transcripts that have been identified primarily by computational means) is likely an undercount because some genes known in other species that are not yet characterized in humans are classified as predicted transcripts, and some genes may remain undiscovered.

In the genome sequence, a small set of genes remain on unplaced chromosome fragments. When predicted transcripts are removed from the total gene count, most of the unplaced genes are eliminated. The RefSeq collection also has an excess of predicted genes, and once these are removed, the remaining difference between the two data sets is rather small.

Difficulties in determining the exact number of protein-coding genes have several sources. Some are accounting issues. The ends of the X and Y chromosomes—known as the pseudoautosomal regions—are essentially identical. For some applications, these are no different from autosomal genes and are counted once. Yet, when the genes are tallied by chromosome, the X and Y chromosomes are considered as distinct chromosomes and, therefore, the pseudoautosomal genes are counted twice. The "– Duplicated Gene Names" column counts the pseudoautosomal genes only once (and removes a few other genes; see the details at the end of this section). Similar issues relate to gene duplications and tandem gene arrays. In some cases, the number of genes in tandem arrays varies among individuals, and recombination of these arrays can produce genes with distinct protein products.

Another problem relates to what is counted as a gene. Although some RNAs that do not encode proteins are labeled as genes on the annotated sequence, familiar RNAs such as transfer RNAs (tRNAs) are labeled as repeats rather than as genes (this is further complicated by the fact that the mitochondrially encoded tRNAs are annotated as genes). In some cases, such as antibodies and T-cell receptors, various gene segments can rearrange to form new genes, and the segments are named separately on the map. Counts in the table on the previous page have been limited to protein-coding genes that do not require rearrangement for expression. The naming of genes causes other complications in counting. For example, the related protein products of the *UGT1A* locus have different names, whereas the unrelated products of the *CDKN2A* locus have the same name. Also, the products of transposons are generally not annotated.

To generate the tables and figures in this volume, different sets of genes were used as appropriate. The genes were counted by name using the appropriate fields in the genome annotation or RefSeq entries. Predicted transcripts and genes were identified by their XM_ and XP_ prefixes. Calculations involving the chromosomes, transcripts, and coding regions typically used the values from the "No. of Mapped Genes" column, with minor adjustments as needed. When a set of mapped transcripts (one per named gene) was collected, the transcripts were selected on the basis of (1) having a larger encoded protein and (2) having a greater number of exons. Pseudoautosomal genes were counted once or twice, as appropriate (see the details at the end of this section). The mitochondrially encoded genes were excluded from figures and other calculations built from the mapped transcripts. Calculations involving protein sequences, in which a connection to the genome was not required, used the RefSeq set (including the mitochondrial genes).

Data Sources, Methods, and References

The pseudoautosomal sequences of the X and Y chromosomes total about 3 Mb, most of which is at the low-numbered coordinates of the reference sequences (short arm). The 17 genes in this larger block, in order, are *PLCXD1* (closest to the end of the short arm), *GTPBP6*, *PPP2R3B*, *SHOX*, *CRLF2*, *CSF2RA*, *IL3RA*, *SLC25A6*, *CXYorf2*, *ASMTL*, *P2RY8*, *DXYS155E*, *ASMT*, *DHRSX*, *ZBED1*, *LOC401577* (predicted), and *CD99*. The genes in the smaller pseudoautosomal region at the ends of the long arms, in order, are *SPRY3*, *SYBL1*, and *IL9R* (closest to the tip of the long arm). There are a large number of additional related gene pairs on the X and Y chromosomes, with varying degrees of sequence divergence and changes in the gene order.

Seven autosomal gene names are duplicated on the map: Four are on different chromosomes and three are on opposite strands of the same chromosome, but only one of each is counted in the table on the previous page. *SKIP* is on chromosomes 2 and 17; *PRG2* is on chromosomes 19 and 11; *DUB3* is on chromosomes 8 and 4; and *LSP1* is on chromosomes 11 and 2. *SLFN5*, *TTMA*, and *ZNF468* are annotated on both strands of the chromosomes where they are located. When predicted genes are removed ("All Genes Minus Predicted Transcripts" row), *TTMA* was eliminated from the gene count (it was predicted on both strands), and discrepancies regarding the placement of *SLFN5*, *LSP1*, and *DUB3* were resolved. Autosomal genes with names that are duplicated on the map were counted once when pseudoautosomal genes were counted once. They were counted twice when pseudoautosomal genes were counted twice.

Is Gene Density Uniform Across the Chromosomes?

No. Gene density is positively associated with guanine + cytosine (G+C) base composition, which varies among the chromosomes.

In the table below, the last column provides an estimate of gene density for each chromosome. The value is lowest for the Y chromosome and highest for chromosome 19.

Chr.	Chr. Size (sequenced Mb)	No. of Genes	Gene Density (genes/sequenced Mb)
1	225.0	2316	10.29
2	237.7	1492	6.28
3	194.7	1184	6.08
4	187.3	905	4.83
5	177.7	1021	5.75
6	167.3	1161	6.94
7	155.0	1079	6.96
8	142.6	797	5.59
9	120.1	900	7.49
10	131.6	907	6.89
11	131.1	1410	10.75
12	130.3	1140	8.75
13	95.6	408	4.27
14	88.3	680	7.70
15	81.3	721	8.86
16	78.9	939	11.90
17	77.8	1260	16.20
18	74.7	341	4.57
19	55.8	1457	26.12
20	59.5	586	9.85
21	34.2	268	7.84
22	34.9	529	15.18
X	151.1	990	6.55
Y	25.7	94	3.66

The following figure shows that gene density correlates well with base composition (G+C content). The points representing chromosomes with extreme gene-density values are labeled.

Values from the last column of the table provide the *x*-axis coordinates for the figure. The *y*-axis coordinates in the figure are from the section on the base composition of the chromosomes (p. 12). The chromosome sizes in the table are based on sequenced regions (see p. 6). The sequenced fraction of the Y chromosome is rela-

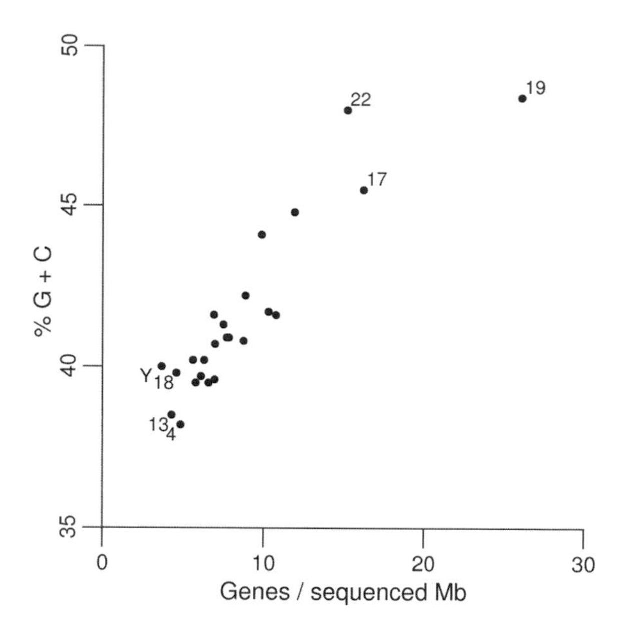

tively low. Therefore, the gene-density value for the Y chromosome would be even lower if the sizes of the intact chromosomes were used.

See page 35 for a related discussion about gene size and gene density.

Data Sources, Methods, and References

The table and figure presented were developed starting with the set of reference transcripts for protein-coding genes in release 36.2 of the human genome sequence, which places 22,585 genes on the chromosomes (see p. 18 for details). Despite the exclusion of a few known genes, this number is likely an overestimate. Gene segments that rearrange to produce other genes (such as those for antibodies and T-cell receptors) are not associated with mapped reference transcripts, so these were excluded from the counts above. Annotated pseudogenes and products of mobile elements were generally excluded as well. Predicted genes were included. They were identified by the XM_ prefix in the name of the transcript. A small number of genes were not placed on the 24 genome sequences and, therefore, were not counted (these were mostly predicted sequences). Genes in the pseudoautosomal regions were counted as present on both the X and Y chromosomes (see p. 19 for details).

How Common Are Pseudogenes?

Release 36.2 of the human genome sequence includes more than 5000 annotated pseudogenes, with a median size of about 1200 nucleotides. They show a complex distribution pattern across the genome. The majority of the annotated pseudogenes are numbered gene predictions. As the genome annotation advances, many additional predicted genes (and small gene fragments that may not be annotated) are likely to be categorized as pseudogenes.

One single-copy gene with an unusual number of pseudogenes is *CYCS* (cytochrome c). The figure below shows the location of the *CYCS* gene and 45 annotated *CYCS* pseudogenes (ψ).

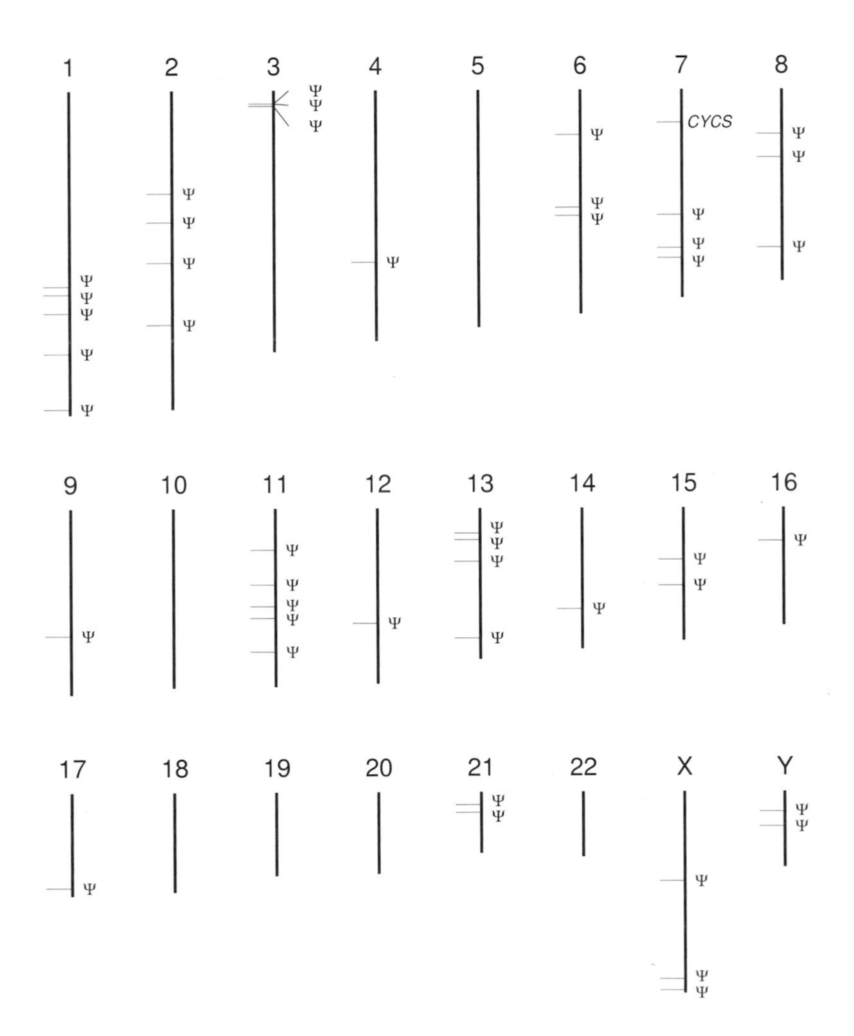

For each functional gene, the number of related pseudogenes is quite variable. A small fraction of annotated pseudogenes are not related to protein-coding genes. Another small class of pseudogenes is involved in rearranging fragments of genes that encode T-cell receptors and antibodies. Pseudogenes are quite common for components of the translation machinery, such as ribosomal proteins. Large gene families, such as the olfactory receptors, also have very large numbers of pseudogenes.

Generally, annotated pseudogenes have a complex pattern of distribution among the chromosomes. Although there is a general positive correlation with chromosome size, several notable exceptions exist. Some exceptions relate to large gene families such as the olfactory and antigen receptors, which add large numbers of small pseudogenes to specific chromosomes.

Gene-rich chromosomes also tend to have more pseudogenes, but exceptions do exist. Of special note is the Y chromosome, which has 176 annotated pseudogenes despite its low gene density (see p. 21). For comparison, chromosome 21 has only 70 pseudogenes but considerably more sequenced DNA.

Additional information on pseudogenes and gene families is found in chapters 7 and 8.

Data Sources, Methods, and References

The *CYCS* and pseudogene locations are from the Map Viewer tables. In addition to *Cycs*, the mouse has a second cytochrome c gene, *Cyct*. One of the human *CYCS* pseudogenes (*HCP9*) is located between *DRB1* (*RBM45*) and *PDE11A* on human chromosome 2. Interestingly, the mouse *Cyct* gene is flanked by their counterparts, *Pde11a* and *Drbp1*, on mouse chromosome 2.

What Is the Size of a Typical Gene?

The median size of protein-coding genes is 16,995 nucleotides, but the sizes of mapped genes vary over a wide range. When transcript predictions are excluded, the median size increases to 21,461 nucleotides. The figure below plots the cumulative size distributions (the percentage of genes up to a given size) for the two sets (50% on the *y*-axis is the median).

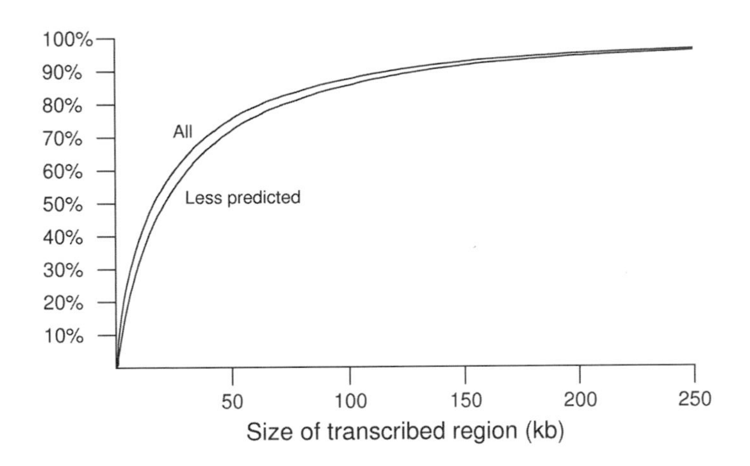

About 4% of genes are more than 250 kb (but these are not plotted here). The very largest genes (p. 26) are far from typical.

With these large sets, there are no clearly defined size classes of genes. In plots of this type, a class of genes of a limited size range would produce a sharp rise at that size.

Data Sources, Methods, and References

The size of a gene is defined here as the genomic span of its unspliced transcript. Only protein-coding genes were included, and for each gene, a transcript for a largest protein was used. In cases in which protein sizes were identical, a transcript with the most exons was retained. Annotated transcripts lacking UTRs were retained here (although in data sets for some of the other figures, they were not). Pseudogenes were excluded. Genes ≥250 kb were not plotted but were used in the totals for normalization.

Which Are the Largest Genes?

Many human genes extend across chromosomal segments that are much larger than needed for their protein-coding regions. The genes with the largest transcribed regions in the human genome are listed below in descending order.

Gene	Gene Size (Mb)	RNA Size (kb)	Protein/Function
CNTNAP2	2.30	9.9	Caspr2 protein
DMD	2.22	14.1	dystrophin
C20orf133	2.06	4.7	
CSMD1	2.06	11.8	
LRP1B	1.90	16.5	lipoprotein receptor family
CTNNA3	1.78	3.0	α-catenin 3
A2BP1	1.69	2.3	ataxin 2 binding protein
FHIT	1.50	1.1	dinucleoside triphosphate hydrolase
GPC5	1.47	2.9	glypican 5
DLG2	1.47	7.7	chapsyn-110
GRID2	1.47	3.0	glutamate receptor
NRXN3	1.46	6.1	neurexin 3
MAGI2	1.44	6.9	membrane guanylate kinase
PARK2	1.38	2.5	parkin
IL1RAPL1	1.37	3.6	receptor accessory protein
CNTN5	1.34	3.9	contactin 5
DAB1	1.25	2.6	*Drosophila* disabled homolog 1
ANKS1B	1.25	4.4	cajalin-2
GALNT17	1.23	3.9	*N*-acetylgalactosaminyltransferase
PRKG1	1.22	3.7	protein kinase
CSMD3	1.21	12.6	
IL1RAPL2	1.20	3.0	receptor accessory protein
AUTS2	1.19	6.0	
DCC	1.19	4.6	netrin receptor
GPC6	1.18	2.8	glypican 6
CDH13	1.17	3.8	cadherin 13
ERBB4	1.16	5.5	EGF receptor family
SGCZ	1.15	2.2	ζ-sarcoglycan
CTNNA2	1.14	3.8	α-catenin 2
SPAG16	1.13	2.2	sperm antigen
OPCML	1.12	6.4	
PTPRT	1.12	12.6	protein tyrosine phosphatase
NRG3	1.11	2.1	neuregulin 3
NRXN1	1.11	6.2	neurexin 1
CDH12	1.10	4.2	cadherin 12
ALS2CR19	1.07	3.5	tight junction protein
PTPRN2	1.05	4.7	protein tyrosine phosphatase
SOX5	1.03	4.5	transcription factor
TCBA1	1.02	3.3	
Genes for Largest Proteins			
TTN	0.28	101.5	titin
MUC16	0.13	43.8	mucin 16

In this table, the size of each gene ("Gene Size" column) is the genomic span of its largest unspliced transcript. The "RNA Size" column shows the size of the corresponding spliced product. For comparison, the gene sizes for the two largest proteins are included. The contrast between the largest genes and the genes for the largest proteins is quite dramatic in terms of the fraction of the gene that is present in the mature RNA.

The chromosomal locations of all of the largest genes listed in the table are shown below. The genes are drawn to scale, with the very largest ones visible as open boxes on the chromosomes. They are widely distributed, but none are present on the chromosomes with the highest gene densities (17, 19, and 22). In a few cases, genes from the same family are linked (e.g., *GPC5* and *GPC6* on chromosome 13).

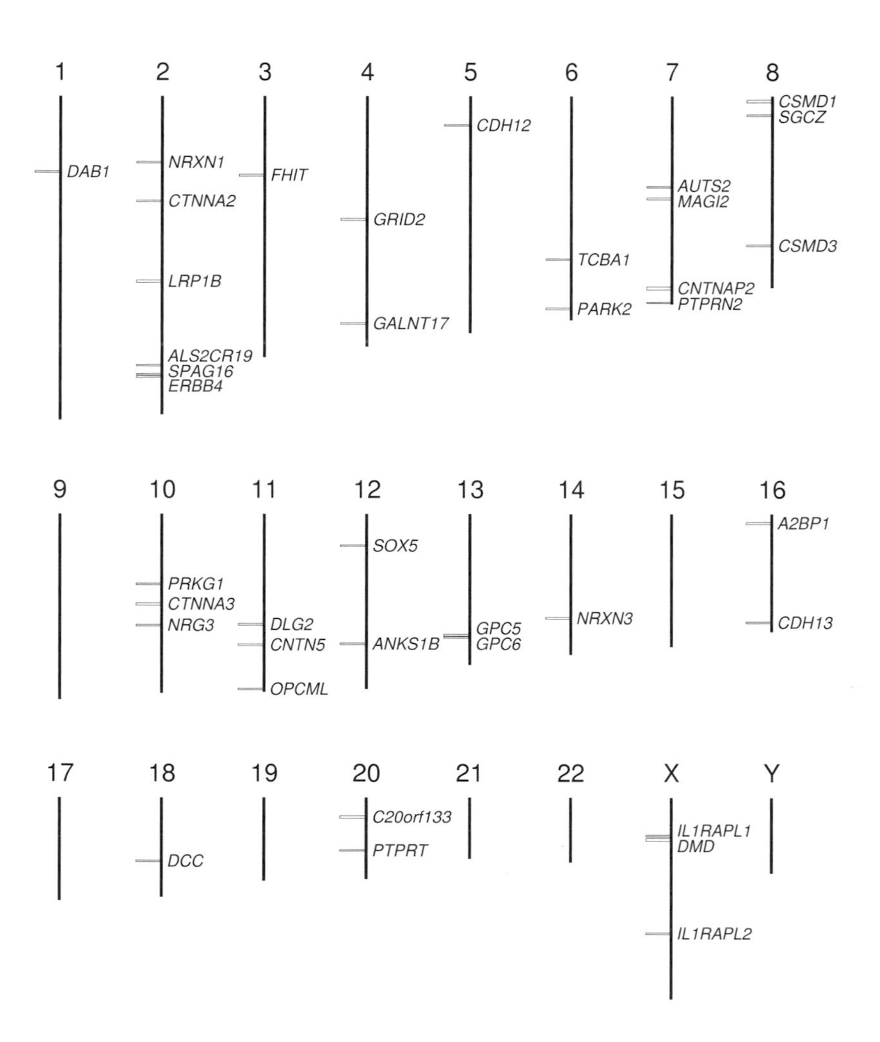

Many of these large genes have functions in the nervous system. Many are members of small gene families, and in some cases, the genes for the other family members are much smaller. For example, *CNTNAP2*, the largest gene (2.30 Mb), is in a family with four other genes that range in size from 0.89 Mb to fewer than 0.02 Mb, but all family members (including *CNTNAP2*) encode similarly sized proteins. Another example is the neurexin gene family: Two neurexin genes (*NRXN3* and *NRXN1*) are listed in the table, and a third family member encodes a similar-sized transcript (compared to those in the table) but is fewer than 0.12 Mb (about one-tenth of the size). A third example is the family of six glypican genes, which produce transcripts in the same size range and encode proteins very close in size. Again, two of the family members are listed in the table (*GPC5* and *GPC6*). The other four glypican family genes range in size from 0.45 Mb to fewer than 0.01 Mb.

Data Sources, Methods, and References

The table and figure were built using the gene location information and chromosomal coordinates in release 36.2 of the human genome reference sequence. All genes greater than 1 Mb are reported, except for *SMA4* (1.04 Mb; the RefSeq entry has been removed) and the hypothetical protein LOC727725 (1.97 Mb; a small protein with many ambiguous positions). Gene sizes were rounded to the nearest 10 kb.

The transcript (RNA) sizes were rounded to the nearest 0.1 kb. In cases in which multiple transcripts spanned the same-sized genomic region, the one with the smaller mature transcript was selected. The transcript sizes may underestimate the 5′ UTR and may include a longer 3′ UTR than is typical for certain genes (see p. 40 for details). For the very large genes, these errors may be significant for the transcript size, but have limited effect on the overall gene size.

Which Genes Have the Most Exons?

The genes with the most exons are presented in the table below. It is important to note that these genes encode many of the largest proteins, but they are generally not among the largest genes.

Gene	No. of Exons	Protein/Function
TTN	312	titin
NEB	150	nebulin
SYNE1	146	nesprin-1
COL7A1	118	collagen type VII
SYNE2	115	nesprin-2
MUC19	108	mucin 19, oligomeric
HMCN1	107	hemicentin 1
RYR1	106	ryanodine receptor (skeletal muscle)
ZUBR1	106	retinoblastoma-associated protein
RYR2	105	ryanodine receptor (cardiac muscle)
RYR3	104	ryanodine receptor
MDN1	102	midasin

Data Sources, Methods, and References

The table was built from the coordinates of the genes and exons mapped onto the reference human genome sequences in release 36.2. All genes with 100 or more exons are presented. For each gene, one transcript (the transcript with the most exons mapped onto the genome) is reported. Some genes might have additional transcripts that have not been mapped onto the genome. In some cases, the total number of exons for that gene may be greater.

Does Exon Number Correlate with Gene Size or Protein Size?

Exon number correlates more with protein size than with gene size, notably for genes with many exons. As described on page 29, the genes with the most exons encode large proteins but are not unusually large genes (see p. 26). The figures below extend this analysis to all genes.

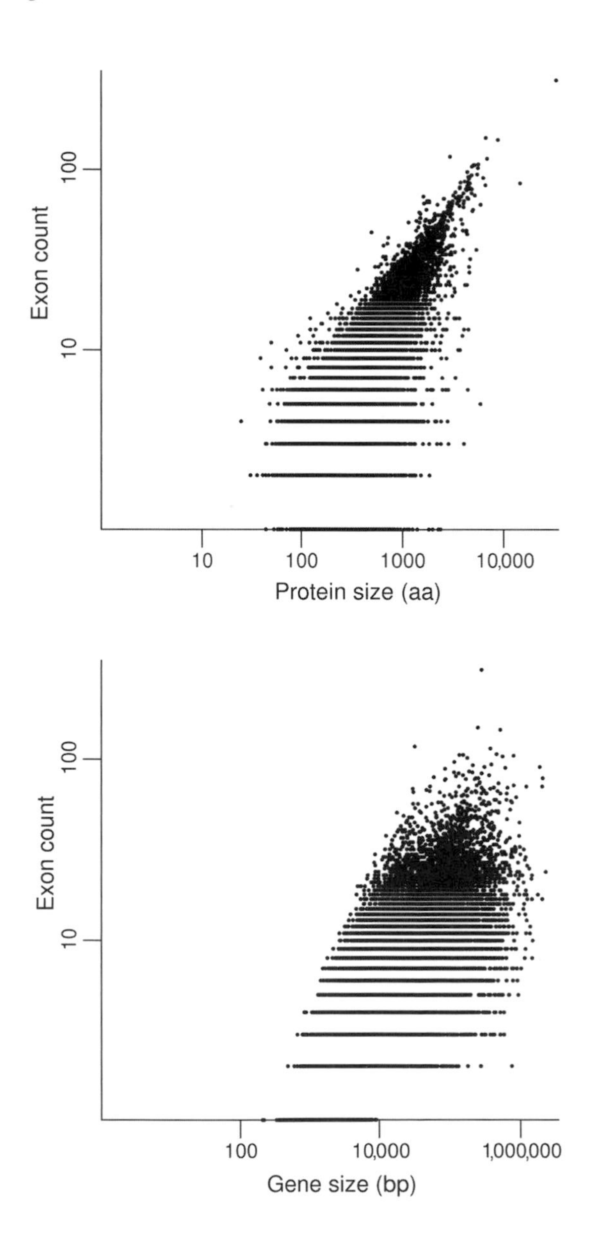

Each point represents one gene, and the points along the x-axes are single-exon genes. Both of the plots are scaled log–log to better show the range of the data. Note the different scales on the x-axes.

Data Sources, Methods, and References

The figures on the previous page are based on the coordinates of the genes and exons mapped onto the reference human genome sequences in release 36.2. The size of each gene corresponds to the genomic span of one of its mapped transcripts (a transcript that encodes for the largest protein). In cases in which alternate transcripts encoded proteins of equal length, a transcript with a greater number of exons was used. Predicted genes were excluded, as were genes without an annotated UTR. (Many genes without UTRs were for olfactory receptors. The UTR lengths of other genes may be underestimated.) Genes from the pseudoautosomal regions of the X and Y chromosomes were counted once. The protein sizes were from the annotated CDS (protein-coding nucleotides including the stop codon, assuming a stop codon was present). For a small number of genes, the annotated CDS was not a multiple of three. In those cases, the protein size was increased by 1 aa.

What Is the Size of a Typical Exon?

Exon sizes are quite variable. Generalizing about them is not appropriate because various categories of exons are quite different in size. The table below presents typical sizes for four classes of exons. For the selected set of transcripts, the median number of exons was 8 and the distribution had a mode of 4.

Type of Exon	Count	Median Size of Exon (bp)	Mean Size of Exon (bp)
Single-exon genes	751	1898	2087
First exon in gene	16,864	181	279
Middle exon in gene	150,672	123	151
Last exon in gene	16,864	941	1325

Middle exons (the largest class of exons) are the smallest. The last exons (3′ ends) are much larger than the first exons (5′ ends). Single-exon genes are typically much larger than the terminal exons of intron-containing genes. In all cases, the mean values were driven by some very large examples, and for the terminal exons, the difference between the means and medians is larger.

It is difficult to establish the sizes of extremely large and small exons. The genome is incompletely annotated, especially with regard to the UTRs (see p. 40 for details). Therefore, the sizes of some reported first exons may be underestimates, or the reported first exons may prove to be internal exons. Some of the last exons may have alternate polyadenylation signals that would produce shorter products.

Data Sources, Methods, and References

The set of transcripts used for this table was also used to produce the figures related to exon counts on page 30 (one transcript per gene was considered; predicted transcripts and genes without UTRs were excluded). All nonterminal exons in genes with three or more exons were classified as middle exons. Means were rounded to whole nucleotides.

See also:

Hawkins J.D. 1988. A survey on intron and exon lengths. *Nucleic Acids Res.* **16:** 9893–9908.

What Is the Size of a Typical Intron?

As described previously, genes show enormous variation in size (p. 25) and in exon number (p. 30). Intron sizes also vary over a wide range, but generalizing across all introns masks differences found in various categories. The table below presents typical sizes for various classes of introns.

Type of Intron	Count	Median Size of Intron (bp)	Mean Size of Intron (bp)
Single-intron genes	1139	2137	8425
First intron in gene	15,725	3549	14,186
Middle intron in gene	134,947	1458	4847
Last intron in gene	15,725	1393	4310

Introns in single-intron genes are intermediate in size between the first and last introns in multi-intron genes. Middle introns (the largest class) are typically much smaller than the first introns of multi-intron genes, but are not notably different from the last introns.

All four intron classes in the table include extremely large and small introns. The largest introns, described on page 34, are much larger than the medians. This results in much larger means for all classes.

Intron sizes show more variation than exon sizes, and the largest introns are much larger than the largest exons.

Data Sources, Methods, and References

The data set used here was identical to the one used for exon sizes (p. 30), but single-exon genes were excluded. Some first introns may, in the future, be found to be middle introns.

See also:

Hawkins J.D. 1988. A survey on intron and exon lengths. *Nucleic Acids Res.* **16:** 9893–9908.

Which Genes Have the Largest Introns?

The table below shows the largest introns that have been mapped in the human genome.

Gene	Size of Largest Intron (bp)	Gene Size (bp)	No. of Introns	Protein/Function
HS6ST3	740,920	748,720	1	heparan sulfate sulfotransferase
GPC5	721,292	1,468,562	7	glypican 5
KCNIP4	693,722	816,041	7	Kv channel interacting protein
SGCZ	682,658	1,148,420	7	ζ-sarcoglycan
CNTNAP2	657,297	2,304,634	23	Caspr2 protein
PCDH9	593,993	927,498	3	protocadherin 9
OPCML	589,253	1,117,529	7	
RORA	550,366	732,040	10	RAR-related receptor
IL1RAPL2	536,480	1,200,827	10	receptor accessory protein
TTC21B	533,197	612,755	29	tetratricopeptide repeat protein
FGF14	526,174	679,090	4	fibroblast growth factor 14
IMMP2L	523,672	899,238	5	inner mitochondrial membrane protease complex
FHIT	522,714	1,502,089	9	dinucleoside triphosphate hydrolase

Of necessity, large introns are found in large genes. The fifth gene listed in the table (*CNTNAP2*) is the largest gene in the genome.

The genes in the table vary considerably in the number of introns present. In some cases, the large intron is only a fraction of the total gene size, while in other cases, it makes up the majority of the gene.

Data Sources, Methods, and References

The information in this table is from the transcripts mapped onto the chromosomes in release 36.2 of the reference genome sequence. Predicted genes have been excluded. The number of introns is from the selected transcript for that gene containing the indicated intron.

Do Gene-rich Chromosomes Have Smaller Genes?

Yes. Gene density varies greatly among the chromosomes (see p. 21). Two factors could be at work to account for the higher gene density on some of the chromosomes: (1) the spacer regions between genes could be smaller or (2) the genes themselves could be smaller. The figure below shows that the gene-rich chromosomes have, on average, smaller genes.

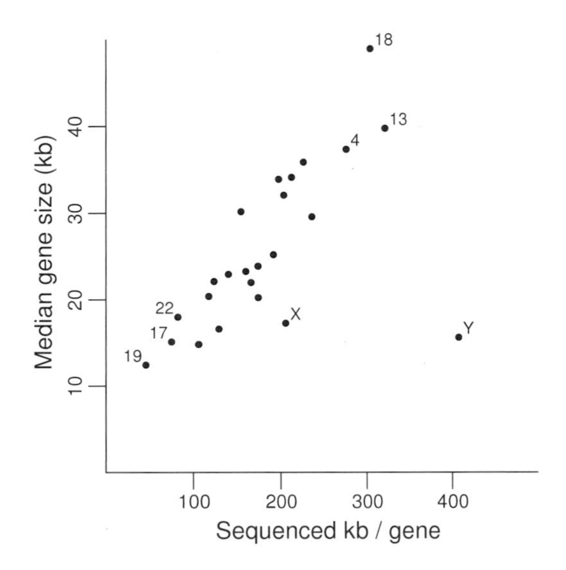

The *x*-axis values (sequenced kb/gene) represent a reciprocal measure of gene density. Chromosomes with extreme values are labeled. Except for the Y chromosome (and to a limited extent the X chromosome), gene density and gene size are highly correlated.

The sums of the transcribed regions of the genes are only fractions of the total sequenced DNA, even on the gene-rich chromosomes (not shown). Changes in spacer regions and gene sizes both contribute to the observed differences in gene density among the chromosomes.

Data Sources, Methods, and References

The gene set used for this figure was also used for the exon number correlations on page 30, except that pseudoautosomal genes were counted as present on both the X and Y chromosomes. To avoid the effects of very large genes, the median gene size was plotted on the *y*-axis. The sequenced region data on the *x*-axis was from the section on chromosome sizes (p. 6). Genes with no reported UTR were excluded (adding them back did not significantly change the results). Gene predictions were also excluded.

See also:

Lander E.S. et al. 2001. Initial sequencing and analysis of the human genome. *Nature* **409:** 860–921.

How Are CpG Islands Associated with Genes?

Distances between the 5′ ends of messenger RNAs (mRNAs) and the nearest CpG islands vary over a wide range, and some are more than 1 Mb. More than 60% of 5′ ends are within 1 kb of a CpG island center (most are much closer). The 5′ ends of most transcripts (59%) are located within the boundaries of an annotated CpG island.

The figure below plots the distances between the 5′ ends (and 3′ ends) of annotated transcripts and the centers of the nearest CpG islands.

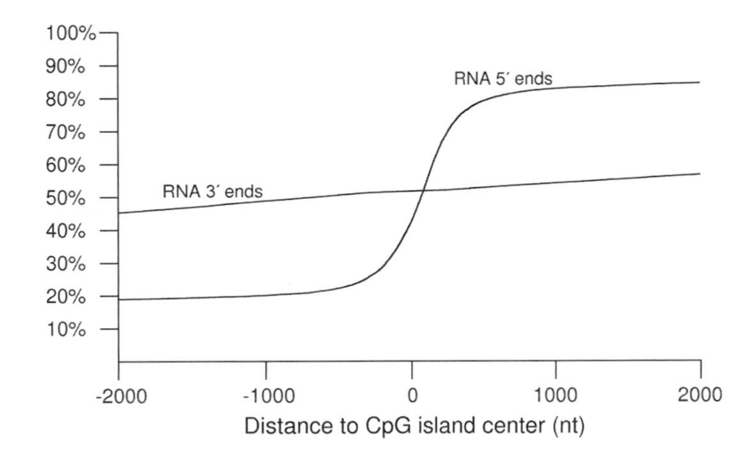

On the x-axis, negative values indicate that the CpG island center is upstream of the 5′ or 3′ end. Only transcripts mapping within 2 kb of the centers of the islands are shown. On the y-axis, 50% is the median distance, and the plot is cumulative.

For comparison, the equivalent data are presented for the 3′ ends of transcripts. Unlike the 5′ ends, the 3′ ends are not necessarily located near CpG islands. Fewer than 3% of the 3′ ends are located in annotated CpG islands.

Data Sources, Methods, and References

CpG islands are regions where the usual underrepresentation of the CpG dinucleotide relative to base composition is less pronounced. CpG islands were from the NCBI Map Viewer annotation for release 36.2 of the genome sequence, and only CpG islands that met their stricter definition were used. There are 24,445 CpG islands (strict definition) annotated on the chromosomes. These have a median size of 960 nucleotides. Collectively, they represent fewer than 1% of the sequenced genome, and are spaced, on average, more than 100 kb apart.

One transcript was selected for each gene, as described on page 31. The closest CpG island was selected using the absolute value of the distance. Among the 17,615 selected transcripts, 13 were on unplaced genomic fragments lacking any CpG islands, and these were excluded. Of the 17,602 usable transcripts, 10,390 had their 5′ ends in an annotated island, and 433 had their 3′ ends in an annotated CpG island. Although only a portion of the complete plot is shown across the *x*-axis, the full data set was used in the percentile calculations for the *y*-axis.

Note that the number of CpG islands is significantly greater than the set of transcripts used. If the calculations are repeated using all transcripts mapped onto the genome (including predicted genes and those with no UTRs, but still excluding those on segments with no CpG islands), similar results are obtained, but the fraction of 5′ ends near CpG islands is somewhat reduced.

Do Large mRNAs Have Large UTRs?

The figures below show how reported 3′ UTR sizes correlate with mRNA sizes and protein sizes.

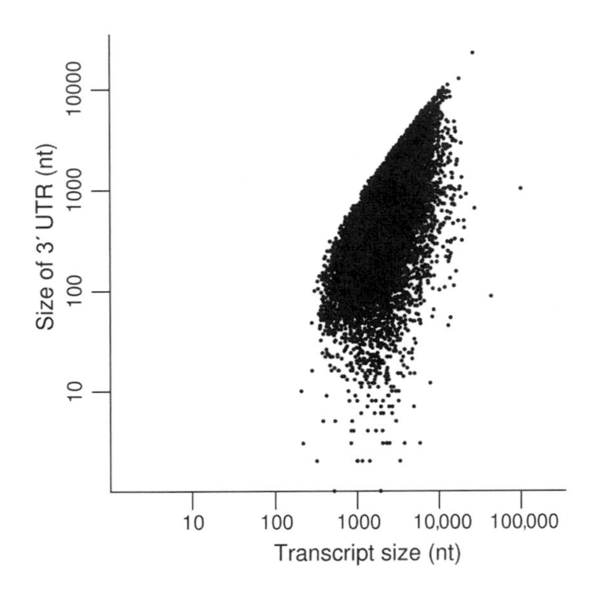

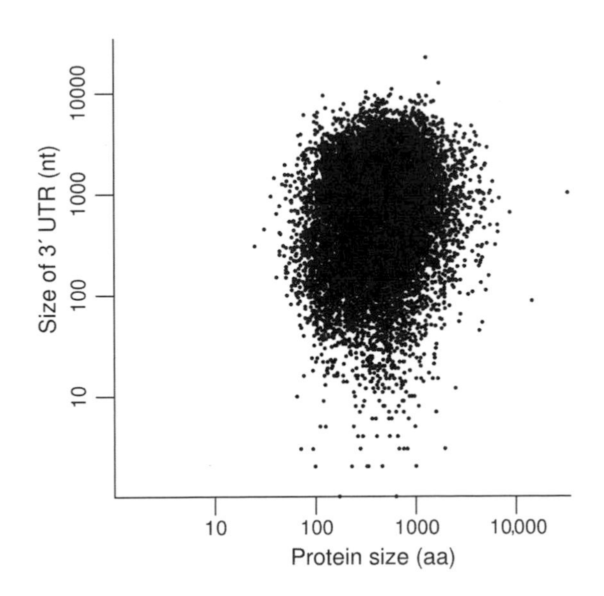

For a significant number of mRNAs, the 3′ UTR is a large fraction of the transcript, thus producing the apparent correlation in the top figure on the previous page (no points can appear above a slope of 1 [note the different scales on the axes] because the mRNA must be longer than its UTR). When one examines how the 3′ UTR sizes are related to protein sizes (bottom figure on the previous page), a strong correlation is not observed. Both plots were scaled log–log to better show the range of the data.

Data Sources, Methods, and References

The initial set of transcripts used in these figures was the same as that used in the figures illustrating correlations with exon number (p. 30). Predicted genes and transcripts with zero reported UTR nucleotides were excluded. Also excluded were transcripts with a 5′ UTR but with no reported 3′ UTR. Genes from the pseudoautosomal regions on the X and Y chromosomes were counted only once.

Accurate calculations related to mRNA size from genome data are difficult. One complication derives from the possibility that complementary DNA (cDNA) sequences are incomplete on the 5′ end. As a result, the 5′ UTRs mapped onto the genome will be similarly incomplete. A second issue is whether known cDNA sequences derive from fully processed transcripts, especially with regard to splicing of introns outside of protein-coding regions. Some well-characterized transcripts are derived from incomplete processing of primary RNAs. Third, some genes are reported with very large 3′ UTRs. These genes may use other polyadenylation signals upstream of the reported 3′ end that would yield smaller RNAs.

Which Genes Are Located in the Introns of Other Genes?

Many genes are found in the introns of other genes. Transcription of these genes is more straightforward when they are encoded on opposite strands of the DNA, or, if encoded on the same strand, when the inner gene contains no introns.

The table below lists some candidates for single-exon genes that are located within an intron of a gene on the same strand. A large number of cases involving genes with limited functional information have been excluded. Only a few of these gene pairs have obvious functional connections.

Inner Gene (transcript size [nt])	Protein/Function	Outer Gene (transcript size [nt])	Protein/Function
CAPNS2 (1009)	calpain subunit	*AYTL1* (74,201)	acetyltransferase family
CHML (7067)	Rab escort protein	*OPN3* (47,250)	encephalopsin
CHMP1B (1272)	ESCRT III complex function	*GNAL* (192,660)	G protein
EID3 (1435)		*TXNRD1* (63,336)	thioredoxin reductase
GALNT4 (5347)	*N*-acetylgalactosaminyl transferase	*WDR51B* (106,277)	
GPR21 (1050)	G-coupled receptor	*RABGAP1* (163,394)	Rab GTPase activator
GPR52 (1472)	G-coupled receptor	*RABGAP1L* (798,694)	
HSN2 (1305)	sensory neuropathy locus	*WNK1* (155,227)	protein kinase
HUS1B (1026)	checkpoint control complex	*EXOC2* (207,972)	exocyst complex
IFRG15 (396)		*TOR1AIP2* (32,982)	torsin-interacting protein
KRTAP3-3 (704)	keratin-associated protein	*KRT40* (21,696)	type I hard keratin family
KTI12 (1696)	transcription	*TXNDC12* (35,245)	thioredoxin family
OMP (492)	olfactory cell function	*CAPN5* (59,210)	calpain 5
OXCT2 (1824)	3-oxoacid CoA transferase 2	*BMP8B* (30,631)	bone morphogenetic protein 8B
P2RY4 (1639)	pyrimidine receptor	*PDZD11* (603,922)	copper metabolism
PPP3R2 (3387)	protein phosphatase regulatory subunit	*GRIN3A* (169,228)	glutamate receptor
PTTG2 (576)	pituitary tumor-transforming family	*TBC1D1* (248,075)	
RNF133 (1443)	ring finger protein }	CADPS2 (567,064)	calcium-dependent exocytosis
RNF148 (1302)	ring finger protein }		
RPA4 (1560)	DNA replication	*DIAPH2* (920,284)	actin polymerization
SLC18A3 (2419)	vesicular acetylcholine transport	*CHAT* (56,010)	choline acetlytransferase
SLC5A3 (2157)	inositol transport	*MRPS6* (69,508)	mitochondrial ribosomal protein
STH (387)	saitohin	*MAPT* (133,924)	microtubule-associated protein
SUMO4 (688)	SUMO modifier family	*MAP3K7IP2* (93,685)	MAP kinase signals
YY2 (1119)	transcription factor	*MBTPS2* (43,120)	transcription factor metalloprotease

The table below lists some candidates for multi-exon genes that are located within an intron of a gene on the same strand. These genes require complex splicing patterns.

Inner Gene (transcript size [nt])	Protein/Function	Outer Gene (transcript size [nt])	Protein/Function
DGAT2L4 (9397)	diacylglycerol acyltransferase family	*PDZD11* (603,922)	copper metabolism
ARHGDIG (2398)	Rho GDP disassociation inhibitor	*PDIA2* (11,599)	protein disulfide isomerase family
CHR415SYT (2704)	synaptotagmin	*TMPRSS11F* (76,672)	serine protease family
SCGB3A1 (1383)	secretoglobin	*FLT4* (63,361)	tyrosine kinase
NUDT22 (3727)	nudix family	*DNAJC4* (10,881)	heat shock protein
MAGEA11 (29031)	melanoma antigen family	*HSFX1* (181,608)	heat shock transcription factor family
KREMEN2 (4165)	Wnt signals	*PAQR4* (11,896)	adiponectin receptor family

Data Sources, Methods, and References

The gene coordinates in release 36.2 of the NCBI reference genome sequence were used to collect candidates for these tables. Only those annotations classified as genes, not pseudogenes, were used. The first step was to identify genes whose coordinates were contained completely within those of another gene. This yielded 648 gene pairs located on the same strand. In addition, 538 annotated gene pairs were located on opposing strands. The lists above are not comprehensive.

Identification of genes within genes is challenging. There are uncertainties about the significance of terminal untranslated exons in the outer gene, as well as uncertainty about missing exons from the inner gene. Of the 648 same-strand gene pairs on the annotated sequence (and the 538 opposite-strand gene pairs), many involved RNAs with no known protein products and were excluded, as were a variety of predicted genes and others with very little functional information. Gene pairs with related products (shared exons) were also excluded, as these transcripts may simply be produced from one gene that undergoes more complex splicing than reported. Similarly, genes representing readthrough transcription products are not shown.

Which Genes Are Present in the Mitochondrial Genome?

The mitochondrial genome encodes 13 protein-coding genes, which are listed with the sizes of their products in the table below. It also has genes for the large ribosomal RNAs (rRNAs) and 22 tRNAs to implement its translation system and distinct genetic code.

Gene	Protein Size (aa)
ND1	318
ND2	347
ND3	115
ND4	459
ND4L	98
ND5	603
ND6	174
CYTB	378
COX1	513
COX2	227
COX3	260
ATP6	226
ATP8	68

All of the protein-coding genes have functions in the respiratory chain. *ND1*, *ND2*, *ND3*, *ND4*, *ND4L*, *ND5*, and *ND6* encode subunits of NADH dehydrogenase. *CYTB* encodes cytochrome B. *COX1*, *COX2*, and *COX3* encode subunits of cytochrome oxidase. *ATP6* and *ATP8* encode F_0 components of the ATP synthase complex.

Data Sources, Methods, and References

The values in the table were derived from the reference mitochondrial sequence GI:17981852. The human mitochondrial genetic code has several differences from the standard code used with nuclear genes (see translation table 2 at http://www.ncbi.nlm.nih.gov/Taxonomy/Utils/wprintgc.cgi?mode=c).

How Are Genes Organized in the Mitochondrial Genome?

The human mitochondrial genome is quite compact. In many ways, it is like a microbial genome. None of the 13 protein-coding genes are annotated with introns.

The following figure shows how the mitochondrial genes are organized.

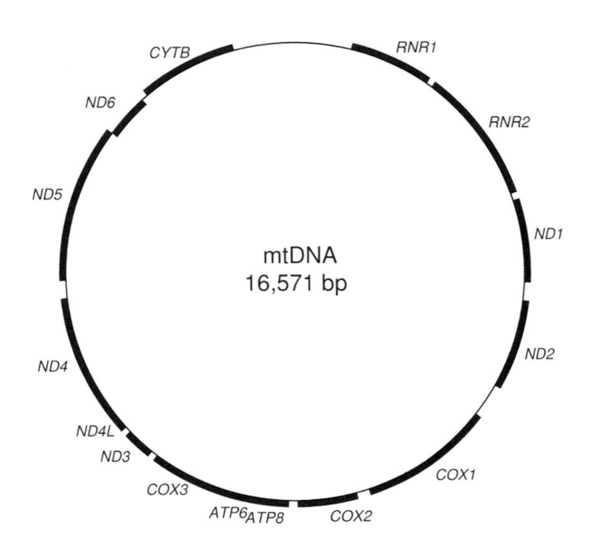

RNR1 and *RNR2* are the main ribosomal RNAs. Except for ND6 (shown on the inside of the circle), all of the genes are transcribed clockwise as shown. Of the 22 mitochondrially encoded tRNAs (not shown), 14 are transcribed clockwise.

Data Sources, Methods, and References

The reference mitochondrial sequence GI:17981852 was used here. The gene coordinates were from the Map Viewer tables. See GI:115315570 for an alternate reference sequence that is two nucleotides shorter.

CHAPTER FOUR

RNA

THERE ARE MANY DIFFERENT MECHANISMS for transcribing and processing human genes. Many of these differences are related to whether the transcript will function as a messenger RNA (mRNA) or as one of the diverse types of nonprotein-coding RNAs (collectively, ncRNAs). Many smaller ncRNAs are encoded in the introns of familiar genes.

The number of ncRNAs with known functions continues to grow. In some cases, the analysis is complicated by the presence of large ncRNA families in the genome. For some of the ncRNAs, most of the copies in the genome are small fragments of the intact genes. Many of the ncRNAs have key roles in processing mRNAs and other ncRNAs.

One of the central characteristics of mRNAs is their exon–intron structure. In this chapter, the sequences used by the major and minor splicing systems are analyzed in some detail.

How Are the Ribosomal RNA Genes Organized in the Genome?

In humans, genes encoding the 18S, 5.8S, and 28S ribosomal RNAs (rRNAs) are located in one transcription unit. These units are tandemly repeated in the five nucleolus organizers on chromosomes 13, 14, 15, 21, and 22. Genes for the 5S rRNA are primarily found in two arrays on chromosome 1.

A prototype 43-kb repeat unit that encodes the 18S (1871 nucleotides), 5.8S (157 nucleotides), and 28S (5035 nucleotides) rRNAs has been assembled, but details about the repeat blocks are missing from the finished genome sequences of the chromosomes. The nucleotide composition of this repeat unit differs greatly from the genome-wide average (see p. 9) and has considerable strand bias.

Nucleotide	Fraction of Repeat Unit (%)
A	14.91
G	26.73
C	31.65
T	26.70

The 5S rRNA is 121 nucleotides. A sequenced repeat from the larger 5S rRNA cluster on chromosome 1 is 2231 nucleotides. Some 5S rRNA repeats are approximately 1.6 kb in length.

Some copy number estimates are presented in the table below. The numbers of repeat units in the tandem arrays are variable. In addition, there are numerous orphans and fragments present in the genome. Additional, more distantly related sequences are also likely to be present.

rRNA	No. of Annotated Segments	No. of Copies in Arrays
18S	17	
5.8S	0	~200 (18S, 5.8S, and 28S together)
28S	93	
5S	1131	~150

Data Sources, Methods, and References

The GenBank rDNA sequence GI:555853 (which contains a repeat unit for the 18S, 5.8S, and 28S rRNAs) was used for the first table on the previous page. For an example of a 5S rDNA repeat unit, see GI:23898.

For the second table, the copy numbers of the arrays were estimates. The numbers of annotated segments were from the reference genome Map Viewer tables. No 5.8S-related sequences were annotated, but two related sequences could be found with BLAST (Mega BLAST with no filter).

See also:

Lander E.S. et al. 2001. Initial sequencing and analysis of the human genome. *Nature* **409:** 860–921.

Sørenson P.D. and Frederickson S. 1991. Characterization of human 5S rRNA genes. *Nucleic Acids Res.* **19:** 4147–4151.

Veiko N.N. et al. 2003. Quantitative analysis of repetitive sequences in human genomic DNA and detection of an elevated ribosomal repeat copy number in patients with schizophrenia (the results of molecular and cytogenetic analysis). *Mol. Biol.* **37:** 409–419.

Worton R.C. et al. 1988. Human ribosomal RNA genes: Orientation of the tandem array and conservation of the 5′ end. *Science* **239:** 64–68.

Which Transfer RNAs Are Present in the Genome?

Determining the number of functional transfer RNA (tRNA) genes in the genome is difficult because of their considerable sequence diversity. The table on the following pages presents examples of genomic sequences for tRNAs with various anticodons.

The sequences shown are trimmed genomic sequences (with introns removed). They have been aligned to show the principal features of the tRNAs, including conserved positions and stem nucleotides. The three adjacent pairs of outward-pointing arrows are the stems of the dihydrouridine (D), anticodon (+++), and pseudouridine (ψ) loops, respectively. The inward-pointing arrows at the ends of each sequence pair form the acceptor stem. Some of these sequences are present as exact matches at more than one location in the genome. Additional related sequences may also be present.

More than one tRNA sequence may exist for the same anticodon. For example, two glutamine tRNAs with the anticodon UUG are listed in the table. The second UUG sequence is related to the human mitochondrial tRNA-Gln sequence. Two methionine tRNAs with the anticodon CAU are listed. The first tRNA-Met is the initiator type, and the second is the elongator type.

Note that a guanine nucleotide is post-transcriptionally added to the 5′ end of tRNA-His. The AUU sequence (for Asn) has some notable differences with the GUU (also Asn) sequence. For a few of the amino acids, the genomic sequence examples differ in length.

Anticodon		Chr.	Size (nt)	Sequence
				------> <--- D loop ---> <---- +++ ----> Variable loop <---- Ψloop ----><------
Ala	UGC	6	72	GGGGGTGTAGCTCAGT-GG--TAGAGCGCATGCTTTGCATGTATGAGGC----------CTCGGGTTCGATCCCCGACACCTCCA
	AGC	6	72	GGGGGTGTAGCTCAGT-GG--TAGAGCGCGTGCTTAGCATGCACGAGGC----------CCCGGGTTCAATCCCCGGCACCTCCA
	CGC	6	72	GGGGATGTAGCTCAGT-GG--TAGAGCGCATGCTTCGCATGTATGAGGC----------CCCGGGTTCGATCCCCGGCATCTCCA
				------> <--- D loop ---> <---- +++ ----> Variable loop <---- Ψloop ----><------
Arg	CCG	17	73	GACCCAGTGGCCTAAT-GG-ATAAGGCATCAGCCTCCGGAGCTGGGGAT----------TGTGGGTTCGAGTCCCATCTGGGTCG
	ACG	14	73	GGGCCAGTGGCGCAAT-GG-ATAACGCGTCTGACTACGGATCAGAAGAT----------TCCAGGTTCGACTCCTGGCTGGCTCG
	UCG	6	73	GACCACGTGGCCTAAT-GG-ATAAGGCGTCTGACTTCGGATCAGAAGAT----------TGAGGGTTCGAATCCCTTCGTGGTTG
	UCU	17	73 + 15	GGCTCTGTGGCGCAAT-GG-ATAGCGCATTGGACTTCTAATTCAAAGGT----------TGTGGGTTCGAATCCCACCAGAGTCG
	CCU	7	73	GCCCCAGTGGCCTAAT-GG-ATAAGGCATTGGCCTCCTAAGCCAGGGAT----------TGTGGGTTCGAGTCCCATCTGGGGTG
				------> <--- D loop ---> <---- +++ ----> Variable loop <---- Ψloop ----><------
Asp	GUC	12	72	TCCTCGTTAGTATAGT-GG-TGAGTATCCCCGCCTGTCACGCGGGAGA-----------CCGGGGTTCGATTCCCCGACGGGGAG
				------> <--- D loop ---> <---- +++ ----> Variable loop <---- Ψloop ----><------
Asn	AUU	1	74	GTCTCTGTGGCGCAATCGG-TCAGAGCGTTCGGCTATTAACCGAACGGT----------GAGTAGTTCAAGACCACCCAGGGACG
	GUU	1	74	GTCTCTGTGGCGCAATCGG-TTAGCGCGTTCGGCTGTTAACCGAAAGGT----------TGGTGGTTCGAGCCCACCCAGGGACG
				------> <--- D loop ---> <---- +++ ----> Variable loop <---- Ψloop ----><------
Cys	GCA	7	72	GGGGGTATAGCTCAGG-GG--TAGAGCATTTGACTGCAGATCAAGAGGT----------CCCTGGTTCAAATCCAGGTGCCCCCT
				------> <--- D loop ---> <---- +++ ----> Variable loop <---- Ψloop ----><------
Glu	CUC	6	72	TCCCTGGTGGTCTAGT-GG-TTAGGATTCGGCGCTCTCACCGCCGCGG-----------CCCGGGTTCGATTCCCGGTCAGGGAA
	UUC	1	72	TCCCTGGTGGTCTAGT-GG-CTAGGATTCGGCGCTTTCACCGCCGCGG-----------CCCGGGTTCGATTCCCGGTCAGGGAA
				------> <--- D loop ---> <---- +++ ----> Variable loop <---- Ψloop ----><------
Gln	CUG	17	72	GGTTCCATGGTGTAAT-GG-TTAGCACTCTGGACTCTGAATCCAGCGA-----------TCCGAGTTCAAATCTCGGTGGAACCT
	UUG	17	72	GGTCCCATGGTGTAAT-GG-TTAGCACTCTGGACTTTGAATCCAGCGA-----------TCCGAGTTCAAATCTCGGTGGGACCT
	UUG	2	72	TAGGACGTGGTGTGATAGG--TAGCATGGAGAATTTTGGATTCTCAGG-----------GATGGGTTCAATTCCTATAGTCCTAG

Continued

Anticodon		Chr.	Size (nt)	Sequence
				------> <--- D loop ---> <---- +++ ----> Variable loop <---- Ψloop ----><------
Gly	CCC	1	71	GCATTGGTGGTTCAGT-GG--TAGAATTCTCGCCTCCCACGCGGGAGA----------CCCGGGTTCAATTCCCGGCCAATGCA
	UCC	19	72	GCGTTGGTGGTATAGT-GG-TTAGCATAGCTGCCTTCCAAGCAGTTGA----------CCCGGGTTCGATTCCCGGCCAACGCA
	GCC	17	71	GCATTGGTGGTTCAGT-GG--TAGAATTCTCGCCTGCCACGCGGGAGG----------CCCGGGTTCGATTCCCGGCCAATGCA
				------> <--- D loop ---> <---- +++ ----> Variable loop <---- Ψloop ----><------
His	GUG	15	72	GCCGTGATCGTATAGT-GG-TTAGTACTCTGCGTTGTGGCCGCAGCAA----------CCTCGGTTCGAATCCGAGTCACGGCA
				------> <--- D loop ---> <---- +++ ----> Variable loop <---- Ψloop ----><------
Ile	AAU	17	74	GGCCGGTTAGCTCAGTTGG-TTAGAGCGTGGTGCTAATAACGCCAAGGT---------CGCGGGTTCGATCCCCGTACGGGCCA
	UAU	19	74 + 19	GCTCCAGTGGCGCAATCGG-TTAGCGCGCGGTACTTATAATGCCGAGGT---------TGTGAGTTCGATCCTCACCTGGAGCA
				------> <--- D loop ---> <---- +++ ----> Variable loop <---- Ψloop ----><------
Leu	UAA	6	83	ACCAGGATGGCCGAGT-GG-TTAAGGCGTTGGACTTAAGATCCAATGGACATATGTCCGCGTGGGTTCGAACCCCACTCCTGGTA
	CAA	6	84 + 24	GTCAGGATGGCCGAGT-GGTCTAAGGCGCCAGACTCAAGTTCTGGTCTCCGTATGGAGGCGTGGGTTCGAATCCCACTTCTGACA
	AAG	16	82	GGTAGCGTGGCCGAGC-GGTCTAAGGCGCTGGATTAAGGCTCCAGTCTCT--TCGGGGGCGTGGGTTCGAATCCCACCGCTGCCA
	UAG	17	82	GGTAGCGTGGCCGAGC-GGTCTAAGGCGCTGGATTTAGGCTCCAGTCTCT--TCGGAGGCGTGGGTTCGAATCCCACCGCTGCCA
	CAG	16	83	GTCAGGATGGCCGAGC-GGTCTAAGGCGCTGCGTTCAGGTCGCAGTCTCC-CCTGGAGGCGTGGGTTCGAATCCCACTTCTGACA
				------> <--- D loop ---> <---- +++ ----> Variable loop <---- Ψloop ----><------
Lys	CUU	16	73	GCCCGGCTAGCTCAGTCGG--TAGAGCATGAGACTCTTAATCTCAGGGT---------CGTGGGTTCGAGCCCCACGTTGGGCG
	UUU	17	73	GCCCGGATAGCTCAGTCGG--TAGAGCATCAGACTTTTAATCTGAGGGT---------CCAGGGTTCAAGTCCCTGTTCGGGCG
				------> <--- D loop ---> <---- +++ ----> Variable loop <---- Ψloop ----><------
Met	CAU	17	72	AGCAGAGTGGCGCAGC-GG--AAGCGTGCTGGGCCCATAACCCAGAGGT---------CGATGGATCGAAACCATCCTCTGCTA
	CAU	8	73	GCCTCGTTAGCGCAGTAGG--TAGCGCGGCAGTCTCATAATCTGAAGGT---------CGTGAGTTCGATCCTCACACGGGGCA
				------> <--- D loop ---> <---- +++ ----> Variable loop <---- Ψloop ----><------
Phe	GAA	19	73	GCCGAAATAGCTCAGTTGG--GAGAGCGTTAGACTGAAGATCTAAAGGT---------CCCTGGTTCGATCCCGGGTTTCGGCA

```
Pro AGG 16 72      ------>  <---  D loop  ---> <----  +++  ----> Variable loop <---- Ψloop ----><------
                   GGCTCGTTGGTCTAGG-GG--TATGATTCTCGCTTAGGGTGCGAGAGGT----------CCCGGGTTCAAATCCCGGACGAGCCC
    CGG 17 72      GGCTCGTTGGTCTAGG-GG--TATGATTCTCGCTTCGGGTGCGAGAGGT----------CCCGGGTTCAAATCCCGGACGAGCCC
    UGG 16 72      GGCTCGTTGGTCTAGG-GG--TATGATTCTCGCTTTGGGTGCGAGAGGT----------CCCGGGTTCAAATCCCGGACGAGCCC

Ser AGA 17 82      ------>  <---  D loop  ---> <----  +++  ----> Variable loop <---- Ψloop ----><------
                   GTAGTCGTGGCCGAGT-GG-TTAAGGCGATGGACTAGAAATCCATTGGGG-TCTCCCCGCGCAGGTTCGAATCCTGCCGACTACG
    UGA  6 82      GTAGTCGTGGCCGAGT-GG-TTAAGGCGATGGACTTGAAATCCATTGGGG-TTTCCCCGCGCAGGTTCGAATCCTGTCGGCTACG
    CGA  6 82      GCTGTGATGGCCGAGT-GG-TTAAGGTGTTGGACTCGAAATCCAATGGGGG-TTCCCCGCGCAGGTTCAAATCCTGCTCACAGCG
    GCU 17 82      GACGAGGTGGCCGAGT-GG-TTAAGGCGATGGACTGCTAATCCATTGTGC-TCTGCACGCGTGGGTTCGAATCCCA-CCTCGTCG

Thr AGU  6 74      ------>  <---  D loop  ---> <----  +++  ----> Variable loop <---- Ψloop ----><------
                   GGCTCCGTGGCTTAGCTGG-TTAAAGCGCCTGTCTAGTAAACAGGAGAT----------CCTGGGTTCGAATCCCAGCGGGGCCT
    UGU 14 73      GGCTCCATAGCTCAGG-GG-TTAGAGCACTGGTCTTGTAAACCAGGGGT----------CGCGAGTTCAAATCTCGCTGGGGCCT
    CGU 17 72      GGCGCGGTGGCCAAGT-GG--TAAGGCGTCGGTCTCGTAAACCGAAGAT----------CGCGGGTTCGAACCCCGTCCGTGCCT

Trp CCA 12 72      ------>  <---  D loop  ---> <----  +++  ----> Variable loop <---- Ψloop ----><------
                   GACCTCGTGGCGCAAC-GG--TAGCGCGTCTGACTCCAGATCAGAAGGC----------TGCGTGTTCGAATCACGTCGGGGTCA

Tyr AUA  2 73 + 20 ------>  <---  D loop  ---> <----  +++  ----> Variable loop <---- Ψloop ----><------
                   CCTTCAATAGTTCAGCTGG--TAGAGCAGAGGACTATAGGTCCTTAGGT----------TGCTGGTTCGATTCCAGCTTGAAGGA
    GUA 14 73 + 21 CCTTCGATAGCTCAGCTGG--TAGAGCGGAGGACTGTAGATCCTTAGGT----------CGCTGGTTCAATTCCGGCTCGAAGGA

Val AAC  6 73      ------>  <---  D loop  ---> <----  +++  ----> Variable loop <---- Ψloop ----><------
                   GTTTCCGTAGTGTAGT-GG-TTATCACGTTCGCCTAACACGCGAAAGGT----------CCCCGGTTCGAAACCGGGCGGAAACA
    UAC 11 73      GGTTCCATAGTGTAGC-GG-TTATCACGTCTGCTTTACACGCAGAAGGT----------CCTGGGTTCGAGCCCCAGTGGAACCA
    CAC  6 73      GTTTCCGTAGTGTAGT-GG-TTATCACGTTCGCCTCACACGCGAAAGGT----------CCCCGGTTCGAAACCGGGCGGAAACA
```

Data Sources, Methods, and References

In Build 36.2 of the reference genome sequence, tRNA-related sequences are annotated as repeats. Some are intact genes identical to known tRNAs, but others are fragments.

The sequences in the table were identified in the genome sequence using BLASTN, starting with annotated tRNA GenBank entries (for human and other vertebrates). For some near-exact matches, the genome sequences had better base pairing in the tRNA stems than did the sequences used to find them.

In all cases, any adjacent CCA sequence in the genome was trimmed. In the "Size" column, the number after the + sign indicates the length of its intron.

See also:

Gu W. et al. 2003. tRNA-His maturation: An essential yeast protein catalyzes addition of a guanine nucleotide to the 5′ end of tRNA-His. *Genes Dev.* **17:** 2889–2901.

Which Genes Host Small Nuclear RNAs and Small Nucleolar RNAs?

The genome has numerous small nuclear RNAs (snRNAs) and small nucleolar RNAs (snoRNAs), both known and predicted, many of which are located within other genes. Many of these are listed, along with their sizes, in the table below.

Several additional small RNAs are antisense to host genes that are listed in the table. These include *SNORD9* in *CHD8*, *SNORA79* in *GTF2A1*, *SNORA22* in *LOC441242*, *SCARNA20* in *USP32*, and *SNORA47* in *ZBED3*. *SNORA66* and *SNORD21* (in *RPL5*) are antisense to the *FAM69A* gene.

Note that numerous ribosomal protein genes (with the *RPL*- and *RPS*- prefixes) function as host genes. A few of the host genes may not encode proteins.

Host Gene (Function)	Size of Host Gene (nt)	snRNA or snoRNA	Size of snRNA or snoRNA (nt)
ATG16L1 (autophagy)	44,016	*SCARNA5*	278
		SCARNA6	266
ATP2B4 (membrane calcium ATPase)	117,282	*SNORA77*	125
ATP5B (mitochondrial ATPase)	7894	*SNORD59A*	75
ATP6V0E (lysosomal ATPase)	51,138	*SNORA74B*	204
BAT2	16,941	*SNORA38*	132
C12orf41	28,805	*SNORA2A*	135
		SNORA2B	137
		SNORA34	137
C17orf45	2987	*SNORD49A*	71
		SNORD49B	48
C6orf160	1655	*SNORD50A*	75
C6orf48	4849	*SNORD48*	64
		SNORD52	64
CCT6A (T complex)	12,305	*SNORA15*	133
CHD3 (chromodomain helicase)	27,398	*SCARNA21*	138
CHD4 (chromodomain helicase)	37,304	*SCARNA11*	137
CHD8 (chromodomain helicase)	51,773	*SNORD8*	109
CNOT1 (CCR4 transcription complex)	109,896	*SNORA46*	135
		SNORA50	136
CWF19L1	35,383	*SNORA12*	147
DKC1 (dyskerin)	14,811	*SNORA36A*	132
		SNORA56	129
EEF1B2 (translation factor)	3336	*SNORA41*	132
		SNORD51	70
EEF2 (translation factor)	9408	*SNORD37*	66
EIF3S10 (translation factor)	45,794	*SNORA19*	128
EIF4A1 (translation factor)	5807	*SNORA48*	135
		SNORA67	137
		SNORD10	148

Continued

Host Gene (Function)	Size of Host Gene (nt)	snRNA or snoRNA	Size of snRNA or snoRNA (nt)
EIF4A2 (translation factor)	6326	*SNORA4*	137
		SNORA63	135
		SNORA81	178
		SNORD2	70
EIF5 (translation factor)	10,870	*SNORA28*	126
EP400 (chromatin remodeling)	130,498	*SNORA49*	137
FAM29A	49,762	*SCARNA8*	131
FLJ10847 (membrane transport)	45,180	*SNORA59B*	152
GAS5	4087	*SNORD44*	61
		SNORD47	77
		SNORD74	72
GNB2L1 (G protein-related)	6979	*SNORD95*	63
		SNORD96A	72
HBII-276HG	3613	*SNORD87*	76
HSPA9B (heat shock protein)	20,092	*SNORD63*	68
HTR2C (serotonin receptor)	326,074	*SNORA35*	128
IPO7 (nuclear transport)	60,872	*SNORA23*	189
JOSD3	11,293	*SNORA1*	130
		SNORA18	132
		SNORA40	127
		SNORA8	139
		SNORD5	73
		SNORD6	71
KIAA0907	21,353	*SCARNA4*	129
		SNORA42	134
KPNA4 (nuclear transport)	65,415	*SCARNA7*	330
LOC128439	4222	*SNORA60*	136
MAGED2 (melanoma antigen family)	8275	*SNORA11*	131
MATR3 (nuclear matrix)	56,339	*SNORA74A*	200
MBD2 (methyl-CpG binding protein)	70,584	*SNORA37*	129
MRPL3 (mitochondrial ribosomal protein)	40,785	*SNORA58*	137
NAP1L4 (nucleosome assembly)	47,948	*SNORA54*	123
NCL (nucleolin)	9747	*SNORA75*	137
		SNORD20	80
NOL5A (nucleolus)	5786	*SNORA51*	132
		SNORD56	71
		SNORD57	72
NOP5/NOP58 (nucleolus)	37,870	*SNORD11*	84
PABPC4 (polyA binding protein)	15,929	*SNORA55*	137
PAR-SN	3183	*SNORD64*	67
PAR5	3182	*SNORD108*	69
		SNORD64	67
PHB2 (mitochondria)	5375	*SCARNA12*	270
POLA (DNA polymerase)	303,037	*SCARNA23*	130
POR (cytochrome P450 reductase)	71,754	*SNORA14A*	134
PPP1R8 (protein phosphatase regulatory subunit)	20,891	*SCARNA1*	166

Continued

Host Gene (Function)	Size of Host Gene (nt)	snRNA or snoRNA	Size of snRNA or snoRNA (nt)
RABGGTB (Rab geranylgeranyl transferase subunit)	8879	*SNORD45A*	84
		SNORD45B	71
RBMX (RNA binding protein)	7321	*SNORD61*	73
RCC1 (regulator of chromosome condensation)	32,935	*SNORA73A*	207
RFWD2 (ubiquitin ligase)	262,404	*SCARNA3*	144
RPL10 (ribosomal protein)	3963	*SNORA70*	135
RPL12 (ribosomal protein)	3730	*SNORA65*	136
RPL13 (ribosomal protein)	2759	*SNORD68*	72
RPL13A (ribosomal protein)	4701	*SNORD32A*	82
		SNORD33	83
		SNORD34	66
		SNORD35A	86
RPL17 (ribosomal protein)	4053	*SNORD58A*	65
		U58	65
RPL18A (ribosomal protein)	3398	*SNORA68*	133
RPL21 (ribosomal protein)	5014	*SNORA27*	126
		SNORD102	72
RPL23 (ribosomal protein)	3733	*SNORA21*	133
RPL23A (ribosomal protein)	4375	*SNORD42A*	62
		SNORD42B	67
		SNORD4A	72
		SNORD4B	74
RPL27A (ribosomal protein)	3091	*SNORA3*	130
		SNORA45	131
RPL3 (ribosomal protein)	6784	*RNU86*	55
		SNORD43	62
		SNORD83A	95
		SNORD83B	93
RPL30 (ribosomal protein)	3832	*SNORA72*	132
RPL37 (ribosomal protein)	3958	*SNORD72*	80
RPL39 (ribosomal protein)	5138	*SNORA69*	132
RPL4 (ribosomal protein)	5532	*SNORD16*	100
		SNORD18A	70
		SNORD18B	70
		SNORD18C	67
RPL5 (ribosomal protein)	9888	*SNORA66*	133
		SNORD21	95
RPLP2 (ribosomal protein)	2941	*SNORA52*	134
RPS11 (ribosomal protein)	3311	*SNORD35B*	88
RPS12 (ribosomal protein)	2996	*SNORA33*	133
		SNORD100	76
		SNORD101	73
RPS13 (ribosomal protein)	3282	*SNORD14A*	92
		SNORD14B	91
RPS2 (ribosomal protein)	2766	*SNORA10*	133
		SNORA64	134

Continued

Host Gene (Function)	Size of Host Gene (nt)	snRNA or snoRNA	Size of snRNA or snoRNA (nt)
RPS20 (ribosomal protein)	1457	*SNORD54*	63
RPS3 (ribosomal protein)	6172	*SNORD15A*	148
		SNORD15B	146
RPS3A (ribosomal protein)	5051	*SNORD73A*	65
RPS8 (ribosomal protein)	3167	*SNORD38A*	71
		SNORD38B	69
		SNORD46	98
		SNORD55	74
RPSA (ribosomal protein)	6121	*SNORA6*	149
		SNORA62	154
SLC25A3 (mitochondrial transport)	8376	*SNORA53*	250
SNRPN (snRNP component)	595,816	*SNORD107*	75
		SNORD108	69
		SNORD109A	67
		SNORD109B	67
		SNORD64	67
SRCAP (chromatin remodeling complexes)	32,476	*SNORA30*	129
TBRG4 (protein kinase-related)	11,619	*SNORA5A*	134
		SNORA5B	132
		SNORA5C	137
TCP1 (T complex)	11,206	*SNORA20*	132
		SNORA29	140
TIGA1	1782	*SNORA13*	133
TNPO2 (nuclear transport)	23,064	*SNORD41*	70
TOMM20 (mitochondrial translocase)	19,472	*SNORA14B*	135
TPT1	3994	*SNORA31*	130
WHSC1	89,422	*SCARNA22*	125

Data Sources, Methods, and References

The data in the table were generated from release 36.2 of the reference human genome sequence. The small RNA genes were generally identifiable by their *SCARNA-*, *SNORA-*, and *SNORD-* prefixes. Small RNA genes located within other genes (excluding pseudogenes and the spacers between rearranging gene segments) were collected.

How Are microRNA Genes Distributed in the Genome?

Annotated microRNAs (miRNAs) have a highly biased distribution on the chromosomes. There are 445 annotated miRNAs in the genome, and they have an average size of 89.5 nucleotides. Of these, 222 are located within other genes (including 24 within pseudogenes). The table below summarizes information about miRNAs within other genes.

miRNA Location	No. of miRNAs Within Genes and Pseudogenes	No. of miRNAs Within Genes Only
Sense strand of host gene	161	160
Antisense strand of host gene	58	35
Both sense and antisense to host genes	3	3
Total	222	198

More often, miRNA genes are located on the sense strand of the host gene. In three cases, miRNAs are located within overlapping genes that are transcribed from opposite strands.

The largest numbers of annotated miRNAs are on chromosome 19 (63), the X chromosome (49), and chromosome 14 (46). Of the 24 miRNAs that are annotated in pseudogenes, 22 are at a single locus on chromosome 19, all in the antisense orientation. After excluding pseudogenes, the largest numbers of miRNAs are on the X chromosome (48), chromosome 14 (46), and chromosome 19 (41).

Data Sources, Methods, and References

This table was constructed using Map Viewer data from release 36.2 of the reference genome sequence. The miRNA genes were identified by their names (beginning with the prefix *MIRN*-). The table only includes genes specifically annotated as miRNAs (other transcripts may yield such products). The coordinates on the reference genome sequence were used to determine whether the miRNA genes were located within host genes. Predicted genes were included as possible host genes in these calculations.

What Is the Size Distribution of snRNA Genes and Related Sequences in the Genome?

The snRNA genes are widely dispersed among the chromosomes. In some cases, they are in tandem arrays. In addition to functional snRNA genes, there are very large numbers of related pseudogenes, fragments, and variants of snRNA genes. These related sequences can arise by premature termination of reverse transcription and subsequent integration into the genome or by other types of DNA rearrangements.

For several of the snRNA genes, most of the related sequences are nearly the same size as the snRNA gene from which they were derived. For example, the size distribution of the U1 sequence family is shown below (note that the tandemly repeated copies of U1 are not included). The U1 RNA is nominally 164 nucleotides. About one-sixth of all annotated U1 sequences are exactly 164 nucleotides, and many others are within a few nucleotides. Of the 187 U1 sequences, 147 are 100 nucleotides or greater in length.

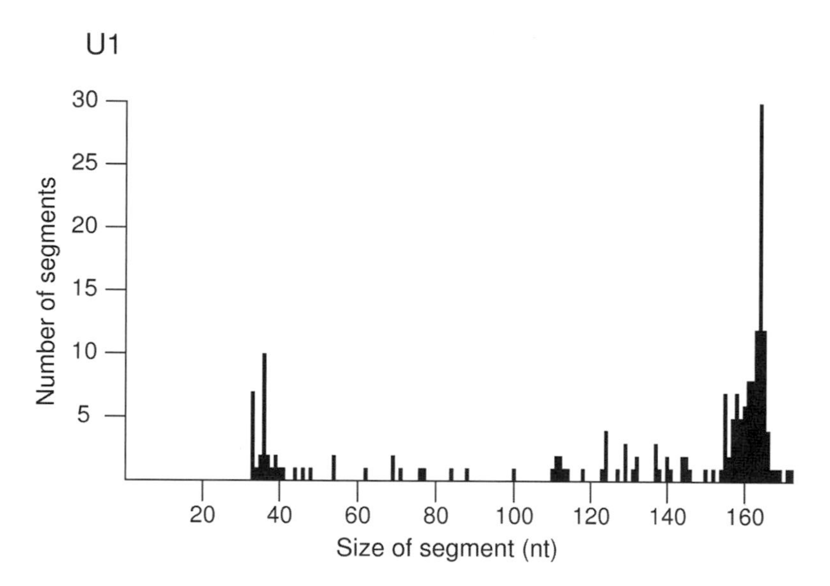

The U6 (1701 segments), U7 (249 segments), U8 (29 segments), and U13 (483 segments) snRNAs have a distribution like U1, with mostly near-full-length sequences present in the genome. Of the 1701 U6 segments in the genome, 203 are 107 nucleotides in length, which is the nominal size of that snRNA.

For other snRNA genes, most of the related sequences are fragments of the snRNA coding sequence from which they were derived. For example, the distribution of U2 sequences is quite different from the pattern seen for the U1 family. First, there are many more copies of U2 sequences in the genome (1075 in total). The vast majority of them are quite small (more like the L1 family of mobile elements described on p. 119). As seen in the following figure, only a small number of the annotated copies are near full length (again, copies in tandem arrays are not included).

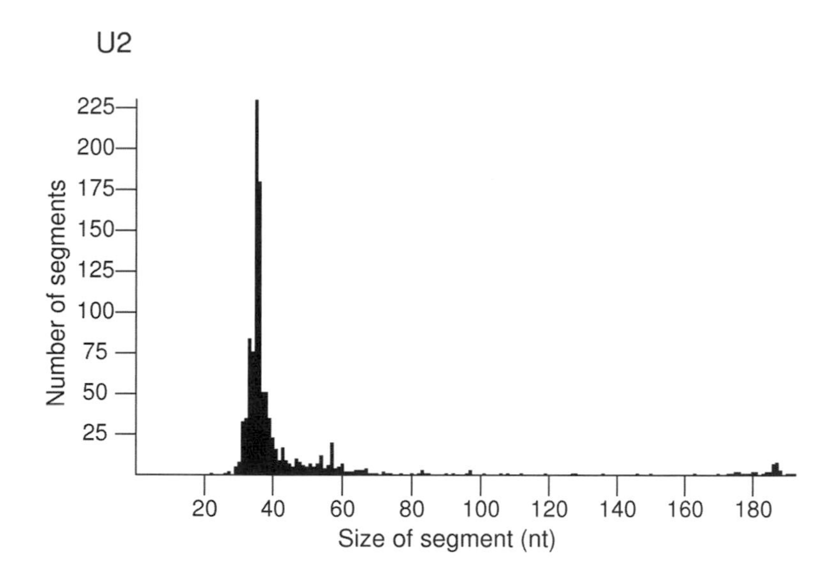

The U3 (228 segments), U4 (164 segments), and U5 (215 segments) snRNAs have a pattern like U2, with mostly fragments of the sequence present. Note that the U7 snRNA is much shorter than the others, so related fragments might not be easily identified. Few sequences related to the U4atac, U6atac, U11, and U12 snRNAs are present in the genome.

Data Sources, Methods, and References

The data for the figures and the other values mentioned in the text are from the table of repeats for the NCBI Map Viewer. Unplaced sequences were not included in the analysis. Adjacent and overlapping segments were merged. The U4atac (GI:3915175), U6atac (GI:1575442), U11 (GI:174934), and U12 (GI:174935) snRNAs were not annotated as repeats on the reference genome sequence and were used in manual searches of assembly 36.2 using BLASTN. For examples of the other sequences, see GI:340085 (U1), GI:1373285 (U2), GI:2736114 (U3), GI:84872049 (U4), GI:84872038 (U5), GI:37562 (U6), GI:174917 (U7), GI:22657578 (U8), and GI:94721317 (U13).

Among other ncRNAs, the 7SL RNA (RN7SL1, GI:84871994) and the 7SK RNA (RN7SK, GI:31455612) both have large families of related sequences in the genome.

See also:

Lander E.S. et al. 2001. Initial sequencing and analysis of the human genome. *Nature* **409:** 860–921.

Li Z. et al. 1998. A tandem array of minimal U1 small nuclear RNA genes is sufficient to generate a new adenovirus type 12-inducible chromosome fragile site. *J. Virol.* **72:** 4205–4211.

What Are the Sequences at Splice Junctions?

About 99% of human introns have the dinucleotide sequences GT and AG at their 5′ and 3′ ends, respectively, and are generally processed by the major spliceosome. A little fewer than 1% of introns have GC and AG, respectively, at these positions. They are also substrates for the major spliceosome. In contrast, a little more than 0.1% of introns have AT and AC at their 5′ and 3′ ends, respectively, and these are generally processed by the minor spliceosome, which can handle variants at the terminal nucleotides (see the notes and references on pp. 62–63). AT–AC introns are found in genes that also have introns that are processed by the major splicing system.

The figure below presents usage data around the 5′ junctions of GT–AG introns (the −1/+1 positions indicate the exon/intron junction). Darker boxes indicate greater usage.

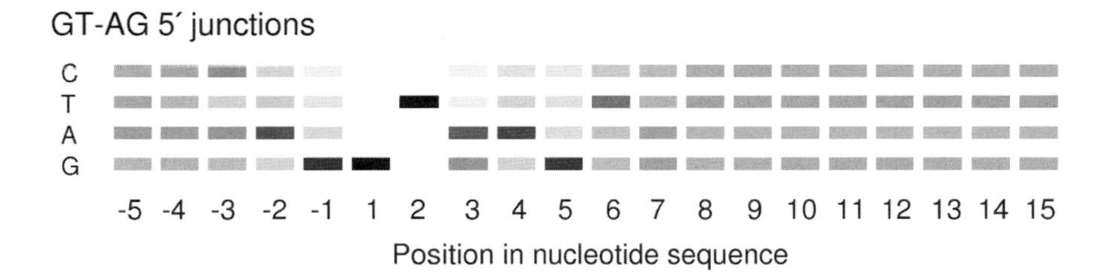

Corresponding data for the GC–AG introns are below. Note the more pronounced biases than seen with the GT–AG introns.

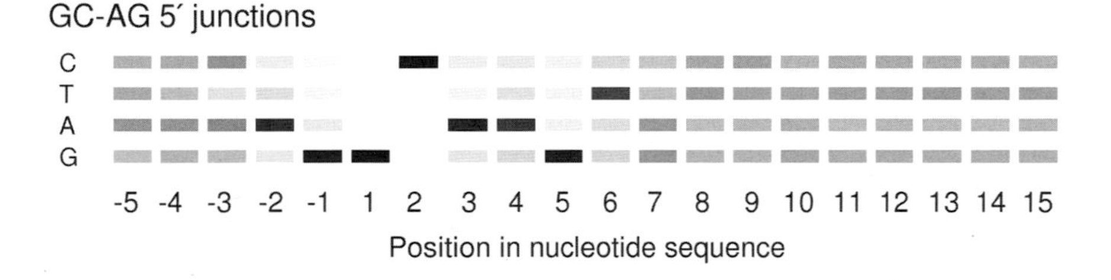

For both the GT–AG and GC–AG introns, strong biases are seen from positions −3 through +6 at the 5′ junctions. At the 3′ junctions, these classes of introns have similar sequences (not shown): The pattern is typically a long pyrimidine tract, no bias at position −4, C > T >> A at −3, an AG dinucleotide at the junction, and a bias toward G at position +1.

At the 5′ junction of AT–AC introns, there is relatively little bias on the exon side, but, as shown in the figure below, very strong bias extends much further into the intron than for GT–AG introns.

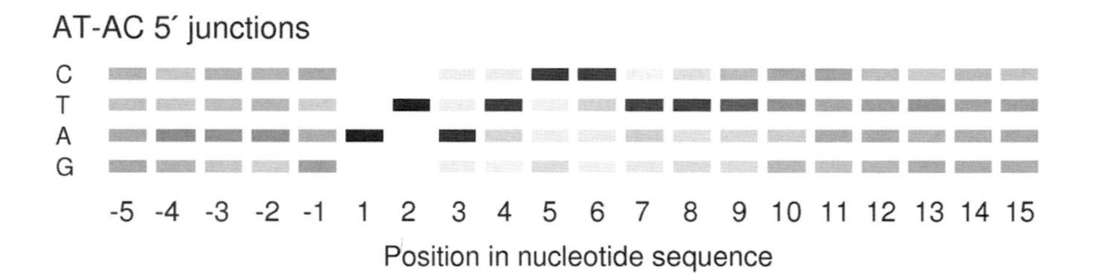

At the 3′ junction of AT–AC introns (shown in the figure below, where the −1/+1 positions indicate the intron/exon junction), the pattern in the intron is generally similar to that for GT–AG and GC–AG introns (except with a C at the last intron position), but the biases are less pronounced. The preference for an A at the first exon position after the intron differs from the G in the other classes of introns.

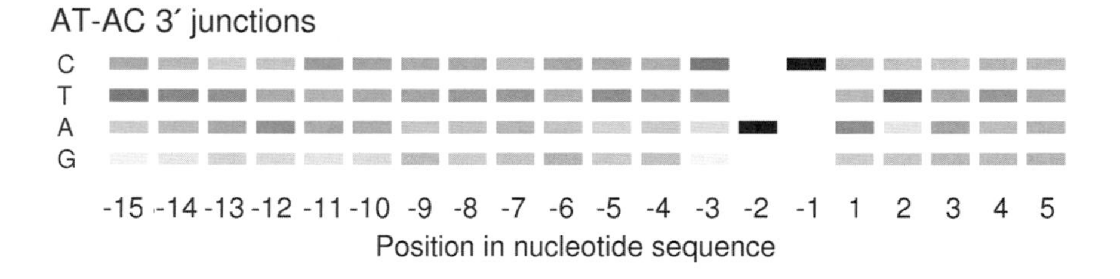

Data Sources, Methods, and References

These figures were built from the transcripts mapped onto Build 36.2 of the human genome sequence. Predicted transcripts were excluded. In cases in which multiple (alternative) transcripts were present for a single gene, identical introns were counted once. When a splice junction was shared with two different partners (in alternate transcripts), both were used because the intact intron was the unit of interest. Introns were categorized solely on the basis of the dinucleotides at the ends of the annotated introns. A total of 173,347 annotated GT–AG introns, 1411 GC–AG introns, and 202 AT–AC introns were used in these calculations. This last class underrepresents the total usage of the minor spliceosome (the minor spliceosome can also process a very small fraction of the GT–AG introns).

See also:

Abril J.F. et al. 2005. Comparison of splice sites in mammals and chicken. *Genome Res.* **15:** 111–119.

Burge C.B. et al. 1998. Evolutionary fates and origins of U12-type introns. *Mol. Cell* **2:** 773–785.

Burset M. et al. 2000. Analysis of canonical and non-canonical splice sites in mammalian genomes. *Nucleic Acids Res.* **28:** 4364–4375.

Which Genes Are Subject to RNA Editing?

For a number of genes, the mature RNA sequences differ from their corresponding genomic sequences because of RNA editing. Two RNA editing mechanisms are described here: one is the conversion of an adenine to an inosine; the other converts a cytosine to a uracil. The table below lists some protein-coding genes where transcripts undergo RNA editing.

Gene	Protein/Function	Editing
APOB[1]	apolipoprotein B	C →U (creates stop codon)
GABRA3[2]	GABA receptor α3	A →I
GRIA2[3]	AMPA-type glutamate receptor	A →I
GRIK2[3]	kainate-type glutamate receptor	A →I (note also *GRIK1*)
HTR2C[4]	serotonin receptor	A →I
IL12RB2[5]	IL12 receptor β2	C →U
KCNA1[6]	potassium channel	A →I
NF1[7]	neurofibromatosis 1	C →U (creates stop codon)

The footnotes are included in "Data Sources, Methods, and References."

Additional loci with adenine-to-inosine RNA editing may include *BLCAP*, *CYFIP2*, *FLNA*, and *IGFBP7*.

Data Sources, Methods, and References

The following footnotes are cited in the table:

[1]Teng B. et al. 1990. Apolipoprotein B messenger RNA editing is developmentally regulated and widely expressed in human tissues. *J. Biol. Chem.* **265:** 20616–20620.

[2]Editing of *GABRA3* has been experimentally validated in other species, but the sequences involved are conserved in the human genome. See:

Ohlson J. et al. 2007. Editing modifies the GABA(A) receptor subunit alpha3. *RNA* **13:** 698–703.

[3]For discussion of glutamate receptors, see:

Kortenbruck G. et al. 2001. RNA editing at the Q/R site for the glutamate receptor subunits GLUR2, GLUR5, and GLUR6 in hippocampus and temporal cortex from epileptic patients. *Neurobiol. Dis.* **8:** 459–468.

Paschen W. et al. 1994. RNA editing of the glutamate receptor subunits GluR2 and GluR6 in human brain tissue. *J. Neurochem.* **63:** 1596–1602.

[4]Flomen R. et al. 2004. Evidence that RNA editing modulates splice site selection in the 5-HT2C receptor gene. *Nucleic Acids Res.* **32:** 2113–2122.

[5]Kondo N. et al. 2004. RNA editing of interleukin-12 receptor beta2, 2451 C-to-U (Ala 604 Val) conversion, associated with atopy. *Clin. Exp. Allergy* **34:** 363–368.

[6]Bhalla T. et al. 2004. Control of human potassium channel inactivation by editing of a small mRNA hairpin. *Nat. Struct. Mol. Biol.* **11:** 950–956.

[7]Skuse G.R. et al. 1996. The neurofibromatosis type I messenger RNA undergoes base-modification RNA editing. *Nucleic. Acids Res.* **24:** 478–485.

See also:

Levanon E.Y. et al. 2005. Evolutionarily conserved human targets of adenosine to inosine RNA editing. *Nucleic Acids Res.* **33:** 1162–1168.

CHAPTER FIVE

PROTEINS

MUCH INFORMATION ABOUT PROTEINS COMES FROM the analysis of complementary DNA (cDNA) and genomic sequences, rather than from the direct analysis of proteins. Although the number of protein-coding genes remains in flux, their main characteristics have largely been determined.

This chapter examines the protein set from two perspectives. One way is to consider the set as a whole and examine their aggregate properties. The second way is to examine specific proteins, such as those with extreme sizes or compositions. In many cases, the unusual composition of a protein is confined to a specific region of the sequence.

Some protein modifications are directed by their terminal sequences. Statistical analyses of the amino and carboxy termini of human proteins are presented.

Many important peptides are processed from larger proteins. In some cases, multiple active peptides are derived from a single precursor. The chapter concludes with some examples of this type.

What Is the Size of a Typical Protein?

The table below provides a summary that includes and excludes predicted proteins.

	All Loci	Excluding Products of Predicted Genes
No. of loci	25,698	18,412
Mean protein size	476 aa	550 aa
Median protein size	341 aa	416 aa
Largest protein	33,423 aa	33,423 aa
Smallest protein	13 aa	25 aa

There are very large proteins in the data set. This leads to mean values that are much greater than the median values. The largest protein is titin, which is encoded by the *TTN* gene. More information about large proteins is presented on page 76.

Excluding predictions, the smallest protein is the product of the ribosomal protein gene *RPL41*. Some predicted proteins are even smaller than this ribosomal protein. The smallest predicted proteins, encoded by *LOC644929* and *LOC727721*, are only 13 aa and are located near each other on chromosome 1.

The figure below presents the size distribution for human proteins fewer than 2000 aa in size, using the data set from the right column of the table above. Given the presence of a few very large proteins, the tail on the right side of the distribution is not surprising.

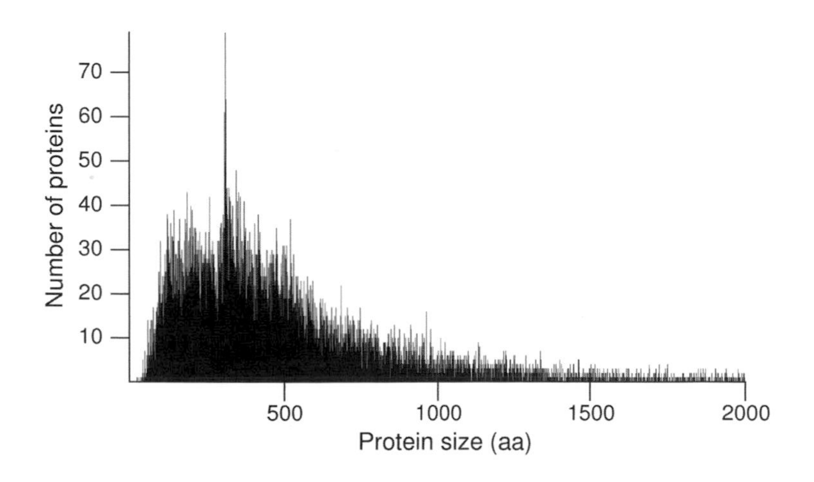

Some of the plot substructure is because of the presence of protein families with members having similar or identical sizes. For example, there is a notable peak on the graph at 312 aa (this is the mode, with 79 proteins; there are also 50 and 55 proteins at 311 aa and 313 aa, respectively). This peak includes a portion of the olfactory receptor family.

Data Sources, Methods, and References

For characteristics such as typical protein size, the choice of data set (e.g., the inclusion of gene predictions) strongly influences the results. Here, the initial data set was 34,410 human RefSeq proteins (available at the time of the release of NCBI reference genome sequence 36.2). Because this set included the products of alternate transcripts, which may encode protein isoforms or identical proteins, a single largest isoform from each named locus was selected (these are reported in the left column of the table on the previous page). More than 28% of these loci were gene predictions. Predicted proteins were identified by the XP_ prefix in the accession numbers (the right column of the table excludes predicted proteins). Proteins $\geq$ 2000 aa were not plotted on the graph. The bin size for the plot was 1.

What Is the Amino Acid Composition of a Typical Protein?

The table below outlines the amino acid composition of human proteins, including those encoded by the mitochondrial genome. The values were generated by averaging (across all proteins) the percent usage of each amino acid. The table also presents the values if the products of predicted genes are excluded.

Amino Acid	Usage (%) (Including Predicted Proteins)	Usage (%) (Excluding Predicted Proteins)
A alanine	7.648	7.197
C cysteine	2.522	2.500
D aspartate	4.267	4.569
E glutamate	6.520	6.756
F phenylalanine	3.546	3.843
G glycine	7.238	6.731
H histidine	2.630	2.603
I isoleucine	3.997	4.385
K lysine	5.477	5.720
L leucine	9.775	10.116
M methionine	2.237	2.295
N asparagine	3.229	3.482
P proline	7.025	6.183
Q glutamine	4.530	4.569
R arginine	6.490	5.822
S serine	8.115	7.951
T threonine	5.158	5.152
U selenocysteine	0.001	0.001
V valine	5.746	6.037
W tryptophan	1.363	1.303
X unknown	0.017	0.001
Y tyrosine	2.468	2.782

If the protein sequences were catenated and the percent usage of each amino acid was then determined, such a calculation would then weight for the composition of larger proteins. Generally, the pattern would be similar to that in the table above, but there would be a decrease in alanine and glycine and an increase in glutamate and aspartate.

The amino acid composition of a protein is generally not uniform along its length. For example, see pages 73–75 for a discussion of the amino acid composition of the amino- and carboxy-terminal regions of proteins.

Data Sources, Methods, and References

The table on the previous page was built from the set of human RefSeq entries available when NCBI Build 36.2 of the genome sequence was released. A largest isoform for each protein was chosen, and its amino acid composition was determined. These compositions were then averaged (across all proteins) to produce the values presented in the table. A total of 25,698 proteins were included in the data set (or 18,412, excluding predictions).

How Do Mitochondrial Proteins Differ in Composition from Typical Proteins?

Compared to the complete set, the 13 mitochondrial proteins are rich in hydrophobic amino acids. Several contain very high fractions of leucine and are some of the most leucine-rich proteins of all (leucine is also typically the highest in all proteins; see p. 70). Mitochondrial proteins are also enriched for isoleucine and methionine, but, except for *ND6*, tend to have reduced valine content.

In the table below, the 13 mitochondrial proteins are listed, and for each protein, the percent usage of selected amino acids is shown.

	Usage (%)						
Protein	F	I	L	M	V	W	Y
ND1	5.0	7.2	19.8	5.0	3.1	2.8	4.4
ND2	4.3	8.9	18.4	7.2	2.3	3.2	2.9
ND3	7.0	7.8	24.3	7.0	2.6	3.5	2.6
ND4	4.4	8.5	20.9	5.9	2.8	2.8	2.8
ND4L	3.1	7.1	23.5	10.2	6.1	0.0	4.1
ND5	6.3	8.8	17.2	4.3	2.7	2.0	2.7
ND6	4.6	6.9	10.9	5.7	17.8	2.9	6.3
CYTB	6.3	10.3	16.9	4.0	2.6	2.9	4.5
COX1	8.0	7.4	12.1	6.2	7.0	3.1	4.3
COX2	4.4	9.7	14.5	4.4	5.7	1.8	4.0
COX3	8.8	5.4	13.1	4.2	5.0	4.6	4.2
ATP6	4.0	12.8	19.5	5.3	3.5	1.3	1.3
ATP8	1.5	4.4	14.7	8.8	1.5	4.4	2.9
Avg. (Mitochondria)	5.2	8.1	17.4	6.0	4.8	2.7	3.6
Avg. (Nucleus & Mitochondria)	3.8	4.4	10.1	2.3	6.0	1.3	2.8

Data Sources, Methods, and References

RefSeq entries were used to calculate the percent usage of each amino acid. Note that some of these proteins are below the size cutoff used in the section on proteins of unusual composition (see p. 80). The averages were computed from the values in the table and were not weighted for the sizes of the proteins.

Which Amino Acids Are Commonly Located in the Amino-terminal Region of a Protein?

In the first position after the initial methionine, the fraction of alanine is greatly elevated. Other notable increases are in glutamate and aspartate. Leucine is the most frequent amino acid overall (see p. 70), but it is greatly reduced in the first position after the initial methionine. Cysteine, histidine, arginine, and tyrosine are among others with significant reductions in the amino-terminal region.

The table below shows the percentage of proteins with a given amino acid at the first three positions after methionine (+1, +2, and +3).

Amino Acid	Usage (%) +1	+2	+3
A alanine	23.05	11.47	9.79
C cysteine	0.95	1.90	1.96
D aspartate	5.59	4.45	3.07
E glutamate	9.46	6.62	6.22
F phenylalanine	1.79	2.77	3.47
G glycine	7.91	7.51	6.92
H histidine	1.05	1.90	1.87
I isoleucine	1.48	2.06	2.92
K lysine	4.30	4.71	4.99
L leucine	5.36	9.42	10.63
M methionine	1.67	1.72	1.65
N asparagine	3.11	2.92	3.02
P proline	5.38	7.33	8.18
Q glutamine	2.34	3.79	4.37
R arginine	4.39	7.34	7.74
S serine	11.58	10.70	9.44
T threonine	4.51	6.02	5.14
V valine	4.04	4.56	5.30
W tryptophan	1.30	1.54	1.57
Y tyrosine	0.74	1.27	1.74

Selenocysteine was not present at positions +1, +2, or +3 in any of the proteins in the data set.

Data Sources, Methods, and References

The set of 18,412 proteins described in the table on page 70 was used here. None of these proteins had amino acid ambiguities at positions +1, +2, or +3.

Which Amino Acids Are Commonly Located in the Carboxy-terminal Region of a Protein?

Amino acid use at the last four positions of a protein is generally similar, except for the final position. At the last amino acid position, the proportions of leucine, valine, and phenylalanine are increased, but the proportions of glutamate and glycine are decreased. The table below shows the percent use of each amino acid for the final four positions (N indicates the last position of the sequence).

Amino Acid	Usage (%) N - 3	N - 2	N - 1	N
A alanine	5.84	6.26	5.94	5.24
C cysteine	2.91	2.25	2.75	3.17
D aspartate	4.52	4.54	5.00	4.86
E glutamate	7.83	6.75	7.13	4.96
F phenylalanine	3.33	3.77	3.52	4.67
G glycine	6.56	6.24	5.95	3.44
H histidine	2.68	2.61	2.82	3.38
I isoleucine	3.98	3.66	3.88	4.48
K lysine	7.62	7.61	7.94	7.53
L leucine	8.64	8.86	8.58	11.12
M methionine	1.93	2.02	1.79	2.24
N asparagine	3.32	3.30	3.76	3.95
P proline	6.49	6.37	5.87	5.54
Q glutamine	4.82	4.59	4.91	4.83
R arginine	6.13	5.44	6.40	5.19
S serine	8.83	10.21	9.35	10.08
T threonine	5.64	6.66	5.36	4.53
U selenocysteine	0.01	0.01	0.02	0.00
V valine	4.92	4.94	4.77	6.61
W tryptophan	1.41	1.32	1.55	1.18
Y tyrosine	2.61	2.61	2.72	3.01

A number of carboxy-terminal protein modification reactions are known. One involves the prenylation of proteins with the CaaX motif ("a" indicating aliphatic) at the carboxy-terminus. In the figure on the following page, the numbers of proteins with cysteine at each of the final 20 positions are shown.

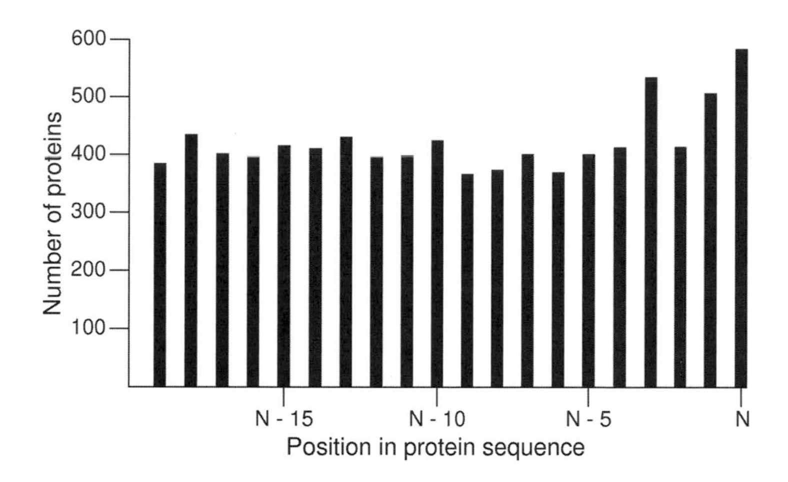

The number of proteins with cysteine at positions N - 19 through N - 4 is relatively constant. At the N - 3 position, there is a noticeable increase. However, the number of proteins with cysteine at N - 3 is exceeded by the number of proteins with cysteine at the final position.

Data Sources, Methods, and References

The complete set of 18,412 proteins reported in the table on page 70 was used here (for the table above and the figure on the following page). None of these proteins had amino acid ambiguities at the last four positions.

See also:

Reid T.S. et al. 2004. Crystallographic analysis of CaaX prenyltransferases complexed with substrates defines rules of protein substrate selectivity. *J. Mol. Biol.* **343:** 417–433.

Which Are the Largest Proteins?

Human proteins extend over a very large size range. The table below presents all genes that encode protein products greater than 5000 aa. A variety of functional types are represented, but many are expressed in muscle.

Gene	Protein Size (aa)	Protein/Function
TTN	33,423	titin
MUC16	14,507	mucin 16 (CA-125)
SYNE1	8797	nesprin 1
MUC19	7328	
SYNE2	6883	nesprin 2
NEB	6669	nebulin
OBSCN	6620	obscurin
GPR98	6306	
C14orf78	6287	
MACF1	5938	filament crosslinking protein
AHNAK	5890	
MUC5B	5765	mucin 5B
HMCN1	5636	hemicentin
MDN1	5596	midasin
DST	5497	dystonin
FCGBP	5405	Fc binding protein
MLL2	5262	
USH2A	5202	usherin
ZUBR1	5183	retinoblastoma-associated protein
MUC2	5197	mucin 2
EPPK1	5065	epiplakin 1
ABCA13	5058	ATP-binding cassette family
RYR1	5038	ryanodine receptor
PCLO	5021	piccolo
KIAA1109	5005	

The genes with the most exons generally produce the largest proteins. Therefore, many of the genes listed in the table on page 29 are also listed here. Many of these genes produce multiple transcripts and protein products, but only the largest isoform is reported in the table. For example, the *TTN* gene produces several smaller protein products. However, given the unusual size of titin, even the smaller products are very large.

Many of the proteins have internal repeated regions. Several of the proteins in the table contain repeats related to those in the spectrin family. However, the spectrins themselves are smaller proteins. The largest of the human spectrins, SPTBN5, is 3674 aa. The others are considerably smaller.

Although some of the proteins in the table are in the mucin family, other mucins are much smaller. RYR1 is in a family with two other members, which are only slightly smaller and did not make the cutoff of 5000 aa for inclusion in the table.

Data Sources, Methods, and References

The table on the previous page was built from the human RefSeq entries that were available when NCBI Build 36.2 of the genome sequence was released. For many loci, the sizes of the RefSeq proteins differ from those annotated on the genome (primarily in cases in which questions remain about the genomic sequence). An extreme example is *EPPK1*: The protein reported on the chromosome map is only 2375 aa, whereas the RefSeq protein is 5065 aa.

Predicted genes were included in this analysis. *MUC19* and *PCLO* (both listed in the table) are predicted genes. Several *PCLO* isoforms have been described. The only current *PCLO* RefSeq isoform is 4935 aa (the *PCLO* isoform listed in the table, which is from the data set that was available when NCBI Build 36.2 of the genome sequence was released, is predicted). LOC727897 (5708 aa) and LOC649768 (5205 aa), which are both predicted, were removed from RefSeq and replaced with the MUC5B protein listed in this table. *KIAA1109* has also been described as the *FSA* gene. *C14orf78* has been named *AHNAK2*.

Because many of these proteins have internal repeated regions, the reference RNAs have not mapped cleanly onto the reference genome sequence. Therefore, the sizes of some of these proteins are likely to be revised in the future.

Which Proteins Have Large Homopolymer Tracts?

Many proteins are notable for having long runs of particular amino acids. Some extreme examples are presented in the table below.

Amino Acid	Examples	Length of Homopolymer Segment (aa)
Alanine	PHOX2B (paired-type homeobox)	20
	FBS1 (fibrosin)	19
	HOXA13 (Hox family homeobox)	18
Arginine	DMRTA2 (related to *Drosophila* doublesex)	12
	SLC24A3 (cation exchanger)	10
Aspartate	HRC (calcium-binding protein)	16
	ATAD2	14
	ASPN (asporin)	14
Glutamate	MYT1 (transcription factor)	32
	EHMT2 (histone methyltransferase)	24
	TTBK1 (tau tubulin kinase)	23
Glutamine	FOXP2 (Fox family transcription factor)	40
	TBP (TATA box binding protein)	38
	MAML2 (Notch signaling pathway)	34
	EP400 (chromatin remodeling)	29
	NCOA3 (histone acetyltransferase)	29
	THAP11	29
	MN1	28
Glycine	AR (androgen receptor)	23
	POU3F2 (POU family transcription factor)	21
	CAPNS1 (calpain subunit)	20
Histidine	FLJ45187	15
	NR4A3 (nuclear receptor family)	14
	DYRK1A (dual-specificity protein kinase)	13
	MEOX2 (homeobox)	13
Serine	SRRM2	42
	MLLT3	42
	ARL6IP4	25
	SETD1A (SET domain)	24
	TNRC18	24
	DACH1 (related to *Drosophila* dachshund)	24

Continued

Amino Acid	Examples	Length of Homopolymer Segment (aa)
Proline	PCLO (piccolo)	22
	FMNL2 (formin family)	21
	ZFHX4 (homeobox family)	20
	WIZ (zinc finger protein)	20
	RAPH1	20
	WHDC1	20
Threonine	IGSF4 (synCAM)	13
	ANK3 (ankyrin family)	12
	JMJD3	11

The codons used in these regions may not be uniform, so the DNA sequences encoding them may not be simple trinucleotide repeats. As a result, the stability of the underlying DNA sequences is variable.

Other proteins have long regions composed primarily of a single amino acid, but they are interrupted by the occasional presence of other amino acids. For this reason, some proteins known for homopolymer segments were excluded from this table. One such example is ATXN1. The polyglutamine region of the ATXN1 reference protein sequence spans 29 aa, but two of the positions have histidines. Again, these substitutions may affect the stabilty of the DNA coding for that part of the protein sequence.

Data Sources, Methods, and References

This table was built from the set of human RefSeq entries obtained at the time of release 36.2 of the genome sequence. Some predicted genes have been excluded. Different cutoffs were selected to provide short lists for each amino acid. The amino acids not shown in the table lacked a small number of notable examples.

Which Proteins Are Rich in Particular Amino Acids?

Many human proteins are especially rich in one or more of the 20 standard amino acids. These proteins carry out a variety of functions and are often members of families. Some examples are listed in the table below.

Amino Acid	Proteins Rich in the Amino Acid	No. of Residues (aa)/Total Length of Protein (aa)*	Abundance of Amino Acid (%)*
Alanine	MARCKS (actin cytoskeleton)	102/332	30.7
	Histone H1 family		
	BASP1	57/227	25.1
	HOXA13 (Hox-type homeobox)	93/388	24.0
Arginine	Arginine/serine-rich splicing factor family		
Asparagine	PYGO1 (related to *Drosophila* pygopus)	50/419	11.9
Aspartate	DSPP (dentin sialophosphoprotein)	259/1301	19.9
	ACRC	122/691	17.7
	SPP1 (osteopontin)	48/314	15.3
	ANP32B	38/251	15.1
Cysteine	Keratin-associated proteins (cysteine-rich type)		
Glutamate	FLJ40113 (golgin family)	126/466	27.0
	RPGR (GTPase regulator)	307/1152	26.6
	ANP32E	71/268	26.5
	NSBP1 (nucleosome-binding protein)	73/282	25.9
Glutamine	IVL (involucrin)	150/585	25.6
Glycine	LOR (loricrin)	145/312	46.5
	NOLA1 (nucleolar protein)	73/217	33.6
	Keratin-associated proteins (high tyrosine and glycine type)		
	Collagens		
Histidine	SLC39A7 (zinc transporter family)	57/429	13.3
	HRC (calcium-binding protein)	89/699	12.7
	HRG	66/525	12.6
Isoleucine	Olfactory receptor families		
	Type 2 taste receptor families		
Leucine	MFSD3	104/412	25.2
	SLC39A5 (zinc transporter family)	123/539	22.8
	GP1BB (platelet glycoprotein Ib subunit)	47/206	22.8
	PLUNC	58/256	22.7
Lysine	Histone H1 family		
	CYLC2 (sperm cytoskeleton)	92/348	26.4
Methionine	RGAG1	145/1388	10.4

Continued

Amino Acid	Proteins Rich in the Amino Acid	No. of Residues (aa)/Total Length of Protein (aa)*	Abundance of Amino Acid (%)*
Phenylalanine	ALG10 (glucosyltransferase)	43/317	13.6
	DERL2 (Derlin family)	31/239	13.0
	DERL3 (Derlin family)	29/235	12.3
	ALG10B	57/473	12.1
Proline	Proline-rich salivary proteins		
Serine	DSPP (dentin sialophosphoprotein)	542/1301	41.7
	HRNR (hornerin)	957/2850	33.6
Threonine	Mucins		
Tryptophan	CCDC70	16/233	6.9
	CDR1	17/262	6.5
Tyrosine	DAZ2 (Daz family)	66/558	11.8
	DAZ3 (Daz family)	46/438	10.5
Valine	PRLHR (prolactin-releasing hormone receptor)	54/370	14.6
	DGCR13	51/353	14.4

*In some cases, protein products from gene families (rather than specific genes) are listed, and no numbers are provided in the last two columns.

For each amino acid, the extreme cases are quite different: LOR contains 46.5% glycine, whereas CCDC70 contains 6.9% tryptophan. Histone H1 proteins are notably high in both alanine and lysine. Additional proteins in the table have high fractions of other amino acids, but are not among the most extreme cases.

HRNR is the largest protein listed in the table (2850 aa, 33.6% of which is serine), but some of the threonine-rich mucins are even larger.

The olfactory receptor family is one of the largest in the genome. Olfactory receptor family members are found not only at the top of the list for isoleucine, but also in classes with slighly lower fractions of isoleucine.

There is very little overlap between the proteins in this table and the one in the section on homopolymer tracts (p. 78).

Data Sources, Methods, and References

This table was built from the set of human RefSeq entries obtained when release 36.2 of the genome sequence became available. For each amino acid, a short list of examples is shown (the cutoffs for inclusion in the list varied by amino acid). Some predicted genes have been excluded. Because of the unusual composition of many small proteins, only examples ≥200 aa are presented in the table. Many of the keratin-associated proteins were shorter than 200 aa, but some were large enough for inclusion. Other proteins, such as the protamines, late cornified envelope proteins, and metallothioneins, were too small to be included in the table.

For Proteins That Are Rich in a Specific Amino Acid, How Are the Residues Distributed Across the Polypeptide Chains?

In some cases, the abundant amino acid is distributed uniformly along the polypeptide chain (e.g., cysteine-rich, keratin-associated proteins), whereas in others, the residues are concentrated in one or a few domains (e.g., threonine-rich mucin 2). The following figures show amino acid distribution patterns for several proteins mentioned on pages 80–81.

RPGR (26.6% glutamate) has a large glutamate- and glycine-rich region. Few other amino acids are present in this segment.

RPGR

E D R K H Y F W M I L V Q N C S T A G P

200 400 600 800 1000

Position in protein sequence

DAZ2 (11.8% tyrosine) has a regular repeating unit that forms much of the protein.

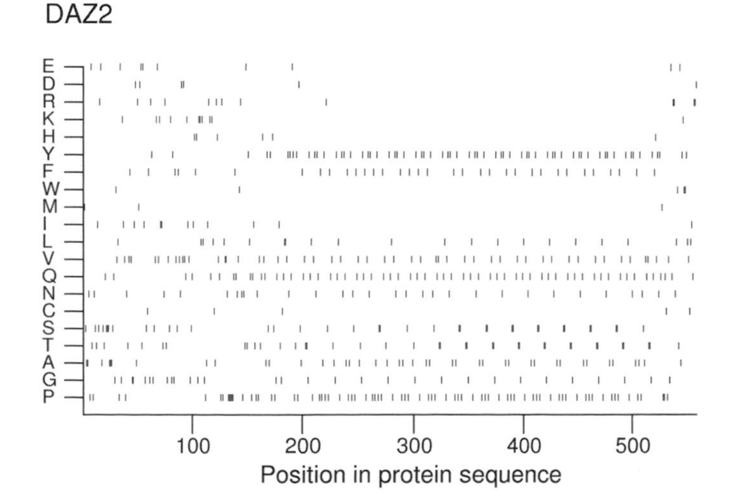

For ALG10 (13.6% phenylalanine) and PRLHR (14.6% valine), the amino acids are largely distributed along the proteins in otherwise complex sequences.

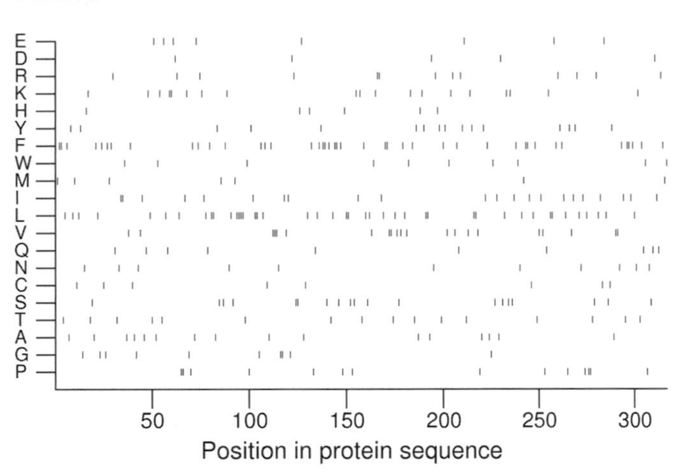

PRLHR

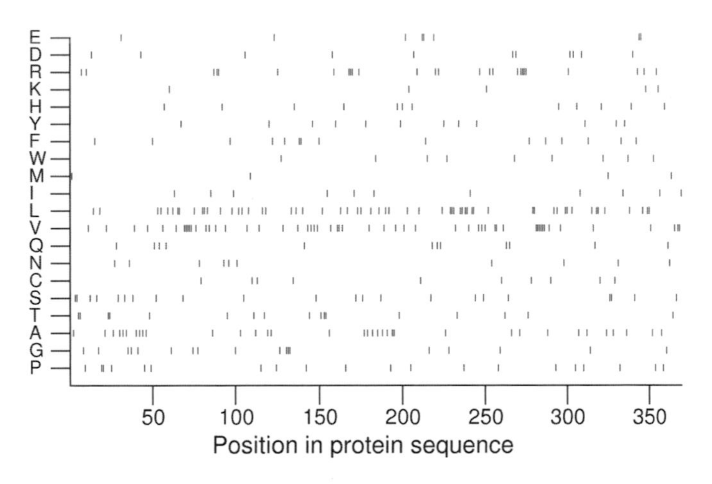

Data Sources, Methods, and References

The following sequences were used to generate the figures: GI:78190492 (for RPGR), GI:11036660 (for DAZ2), GI:44680156 (for ALG10), and GI:4758474 (for PRLHR).

Which Proteins Lack Specific Amino Acids?

Many proteins lack one or more of the 20 standard amino acids. In most cases, these proteins are small or have a large region of low complexity, so it is not surprising that an amino acid is absent, simply by chance. The table below shows the largest proteins that lack each of the amino acids (or in the case of methionine, have only the single residue at the start).

Amino Acid	Proteins that Lack the Amino Acid	Length of Protein (aa)*
Alanine	Keratin-associated proteins	
	NDUFB6 (NADH dehydrogenase subunit)	128
	LCE1 family	
Arginine	LOR (loricrin)	312
	Keratin-associated proteins	
	MS4A8B	250
	BASP1	227
Asparagine	ELN (elastin)	757
	Keratin-associated proteins	
	RHBDD3	386
Aspartate	Keratin-associated proteins	
	CYB561D1 (cytochrome B561-related)	229
	CYB561D2 (cytochrome B561-related)	222
Cysteine	CCDC13	715
	NOLC1 (nucleolar protein)	699
	UBC (ubiquitin C)	685
	WBP11	641
	UBQLN2 (ubiquilin 2)	624
	SCG2 (chromogranin C)	617
	UBQLN1 (ubiquilin 1)	589
Glutamate	LOR (loricrin)	312
	Keratin-associated proteins	
Glutamine	CRIPAK (protein kinase inhibitor)	446
	CDR1	262
	FAIM	213
Glycine	CST11 (cystatin 11)	138
Histidine	Keratin-associated proteins	
	PAWR	340
	SLC35A1 (CMP-sialic acid transporter)	337
Isoleucine	MARCKS	332
	AQP12A (aquaporin 12A)	295
	APOA1 (apolipoprotein A1)	267
Leucine	LOR (loricrin)	312
	Keratin-associated proteins	

Continued

Amino Acid	Proteins that Lack the Amino Acid	Length of Protein (aa)*
Lysine	GPR152 (G-coupled receptor)	470
	RTN4RL2 (Nogo receptor family)	420
	GDF1 (bone morphogenetic protein family)	372
Methionine	ANKRD56 (ankyrin repeat domain family)	793
	ELN (elastin)	757
	DDN (dendrin)	657
Phenylalanine	Keratin-associated proteins	
	PRB family	
Proline	Tropomyosins	
Serine	NDUFA5 (NADH dehydrogenase subunit)	116
Threonine	PRB2 (proline-rich salivary protein)	331
	PRB1 (proline-rich salivary protein)	331
	Keratin-associated proteins	
Tryptophan	TAF4 (TFIID component)	1085
	SPINK5 (serine protease inhibitor)	1064
	BICD1 (related to *Drosophila* bicaudal D)	975
	KM-HN-1	833
	MCM3 (MCM DNA replication complex subunit)	808
Tyrosine	POLR3D (RNA polymerase III subunit)	398
	RASSF7	373
	PSRC1	363
	MDFIC	355
	PRB3 (proline-rich salivary protein)	351
Valine	NGFRAP1	111

*In some cases, protein products from gene families (rather than specific genes) are listed, and no numbers are provided in the last column.

Many of the proteins listed above do contain large low-complexity regions. Several proteins lack more than one amino acid (e.g., the LOR protein lacks arginine, glutamate, and leucine, as well as asparagine [not shown in table]). For amino acids that are less common overall (see p. 70), the proteins in the table tend to be larger in size. For example, tryptophan is the least-frequent amino acid (it has an overall frequency of 1.3%), and it has the largest proteins in the list (>1000 aa).

Data Sources, Methods, and References

The set of proteins used here excludes predicted proteins (the same ones used for the right column of the table on p. 70 were used). For each amino acid, a short list of examples is shown (the cutoffs for inclusion in the list varied by amino acid). The last column ("Length of Protein") indicates the size of the primary translation product. Some proteins were larger than those listed but had limited functional information so were omitted from the table. These were C14orf155, C6orf111, and C10orf68 (for cytosine); LOC284861, LOC285908, and FLJ44790 (for aspartate); LOC388022 and FLJ35934 (for glutamate); LOC387646 and FLJ43339 (for phenylalanine); LOC284861, C10orf96, FLJ35934, FLJ45079, FAM33A, and CXorf1 (for glycine); FLJ46309, LOC389816, and LOC388022 (for isoleucine); LOC402110 and FLJ46309 (for lysine); LOC388438 (for asparagine); LOC388022 and LOC284861 (for glutamine); LOC284861, C10orf95, and FAM27E1 (for valine); KIAA0240 (for tryptophan); and LOC388438, FLJ46309, and LOC144097 (for tyrosine).

Which Proteins Are Post-translationally Cleaved into Multiple Hormones and Related Peptides?

Some human genes encode proteins that are post-translationally cleaved into more than one small peptide hormone. Sometimes, the cleavage process generates intermediates or alternate products that are also biologically active. The table below focuses on genes that encode nonoverlapping protein products (not all protein products are listed).

Gene	Products of Post-translational Cleavage
AVP	arginine vasopressin, neurophysin II, copeptin
GCG	glicentin-related, glucagon, glucagon-like 1, glucagon-like 2
GHRL	ghrelin, obestatin
HCRT	orexin A, orexin B
NTS	neuromedin N, neurotensin
OXT	oxytocin, neurophysin I
PDYN	β-neoendorphin, dynorphin, leumorphin, rimorphin, leu-enkephalin
PENK	met-enkephalin (four copies), met-enkephalin (two extended types), leu-enkephalin
POMC	ACTH, melanotropins, lipotropins, β-endorphin
TAC1	substance P, neuropeptide K, neurokinin A

Except for *PENK* and *POMC*, the peptides are listed in the order in which they appear in the polypeptide precursor. *POMC* and *TAC1* produce many protein products via alternative processing. Note also the products of alternative splicing at *TAC4* (*TAC4* is a distant relative of *TAC1*).

Data Sources, Methods, and References

Most of the peptides listed in the table are annotated on the RefSeq entries.

See also:

Legon S. et al. 1982. The structure and expression of the preproenkephalin gene. *Nucleic Acids Res.* **10:** 7905–7918.

Page N.M. et al. 2003. Characterization of the endokinins: Human tachykinins with cardiovascular activity. *Proc. Natl. Acad. Sci.* **100:** 6245–6250.

Seim I. et al. 2007. Revised genomic structure of the human ghrelin gene and identification of novel exons, alternative splice variants and natural antisense transcripts. *BMC Genomics* **8:** 298.

CHAPTER SIX

TRANSLATION AND PROTEIN MODIFICATION

HUMAN NUCLEAR GENES USE THE STANDARD GENETIC CODE. In addition, a small set of genes uses UGA codons to encode for selenocysteine. The 13 genes of the mitochondrial genome (see p. 43) are translated with a different genetic code.

Individual genes can vary greatly in their use of synonomous codons for a given amino acid. These differences can be seen among members of the same gene family.

Proteins may contain amino acids that are produced by modification of standard amino acids after translation. Some of these modifications are functionally important. They may be required for activity or may serve a regulatory role. Two of these modifications, the conversion of cysteine to formylglycine and the conversion of glutamate to γ-carboxyglutamate, are presented in more detail.

How Does Codon Usage Vary Among Human Genes?

There is considerable variation in the use of synonymous codons among human genes. Efficiently translated genes often show considerable bias in these choices. Even within gene families, members can show very different patterns of codon usage. In the table below, the codon usage patterns for two similarly sized members of the tubulin family are detailed. TUBB2A, a 445-aa protein, is one of the β types; TUBD1, a 453-aa protein, is the single δ type.

Codon	Amino Acid	TUBB2A	TUBD1	Codon	Amino Acid	TUBB2A	TUBD1	Codon	Amino Acid	TUBB2A	TUBD1	Codon	Amino Acid	TUBB2A	TUBD1
UUU	F	1	12	UCU	S	4	14	UAU	Y	2	7	UGU	C	2	5
UUC	F	22	11	UCC	S	6	5	UAC	Y	14	9	UGC	C	5	3
UUA	L	0	8	UCA	S	3	12	UAA	*	1	0	UGA	*	0	1
UUG	L	1	9	UCG	S	3	0	UAG	*	0	0	UGG	W	4	8
CUU	L	0	10	CCU	P	3	9	CAU	H	2	12	CGU	R	0	2
CUC	L	7	8	CCC	P	12	3	CAC	H	8	4	CGC	R	12	0
CUA	L	0	1	CCA	P	2	4	CAA	Q	1	10	CGA	R	0	2
CUG	L	24	6	CCG	P	3	0	CAG	Q	21	15	CGG	R	6	3
AUU	I	1	14	ACU	T	0	5	AAU	N	2	15	AGU	S	3	12
AUC	I	18	9	ACC	T	17	2	AAC	N	21	12	AGC	S	11	4
AUA	I	0	3	ACA	T	3	6	AAA	K	1	15	AGA	R	2	4
AUG	M	19	17	ACG	T	9	1	AAG	K	14	11	AGG	R	2	2
GUU	V	0	15	GCU	A	4	10	GAU	D	5	9	GGU	G	1	8
GUC	V	7	3	GCC	A	19	7	GAC	D	21	8	GGC	G	22	4
GUA	V	1	3	GCA	A	1	12	GAA	E	4	13	GGA	G	4	13
GUG	V	21	7	GCG	A	4	0	GAG	E	32	14	GGG	G	8	3

*Indicates a stop codon. For the 1-letter amino acid code, see page 70.

There are many examples of synonymous codons in the table above where a G or a C is preferred at the third base of the codon in *TUBB2A* but a more uniform use or a slight bias in the other direction is seen with the *TUBD1* gene.

Data Sources, Methods, and References

The table above was constructed using the translated regions in the GenBank entries GI:68299771 (for *TUBB2A*) and GI:50592997 (for *TUBD1*).

Which Proteins Contain Selenocysteine?

Human proteins may contain a variety of amino acids besides the standard 20. Although most nonstandard amino acids are created by post-translational modifications (see p. 93), selenocysteine (abbreviated U or Sec) is incorporated into a small number of proteins by a special transfer RNA (tRNA) that is used at specific UGA codons. The table below lists proteins that contain selenocysteine.

Gene	Protein Product	Notable Characteristics
C11orf31	selenoprotein H	
DIO1	thyroid deiodinase 1	
DIO2	thyroid deiodinase 2	contains two or three selenocysteines (depending on the isoform)
DIO3	thyroid deiodinase 3	
GPX1	glutathione peroxidase 1	
GPX2	glutathione peroxidase 2	
GPX3	glutathione peroxidase 3	
GPX4	phospholipid hydroperoxidase	
GPX6	glutathione peroxidase 6	
SELI	CDP-ethanolamine ethanol-aminephosphotransferase	related to *CEPT1*
SELK	selenoprotein K	related to proteins in vertebrates and sea urchins
SELM	selenoprotein M	
SELO	selenoprotein O	related to proteins in diverse eukaryotes but not in *C. elegans* or *D. melanogaster*
SELS	selenoprotein S	
SELT	selenoprotein T	related to proteins in diverse eukaryotes
SELV	selenoprotein V	proline-rich
SELW1	selenoprotein W1	related to SELV
SELX1	methionine sulfoxide reductase	
SEP15	15-kDa selenoprotein	related to proteins in diverse eukaryotes
SEPHS2	selenophosphate synthetase 2	involved in the selenium utilization pathway
SEPN1	selenoprotein N	related to proteins in vertebrates and sea urchins; contains one or two selenocysteines (depending on the isoform)
SEPP1	selenoprotein P	contains 10 selenocysteines
TXNRD1	thioredoxin reductase 1	also related *TXNRD3* (see the notes at the end of this section)
TXNRD2	thioredoxin reductase 2	

Many of the protein products are quite small (more than a third are annotated as having fewer than 200 aa). A number of selenocysteine residues are located close to the carboxy-termini of the proteins. The *SEPP1* gene encodes the largest number of selenocysteines (ten).

For some of the enzymes, the selenocysteine residues are located within their active sites. Interestingly, the product of *SEPHS2* is involved in the selenium utilization pathway that synthesizes selenocysteine.

Some genes that encode selenocysteine are members of small families. In the table on the previous page, the presence of related proteins in other species is noted (these may or may not be selenoproteins).

Data Sources, Methods, and References

The table on the previous page was built from the set of human RefSeq entries available when release 36.2 of the genome sequence became available. There are a considerable number of ambiguous residues in the RefSeq protein sequences. All proteins with a U are reported in this table.

There may be additional selenoproteins other than those reported in the table: Some predicted proteins may encode selenocysteines, but, in the current annotation, are reported to have UGA stop codons. One such candidate is *TXNRD3*, a member of the thioredoxin reductase family, whose annotation ends with a UGA stop codon (but a selenocysteine is present in the corresponding position in other family members).

For the search results reported in the "Notable Characteristics" column, the human proteins were used as queries in BLASTP searches of RefSeq proteins from human and other sequenced genomes. Some of the human selenoproteins are small, so related proteins may not have been detected in these simple BLASTP searches.

The RefSeq record for *GNRHR2* was not included in this set, even though this gene was previously thought to encode a selenocysteine. The human *SMCP* gene does not contain a selenocysteine. However, a mouse gene named *Smcp* is reported to contain a selenocysteine. Both of these genes encode relatively small, cysteine-rich proteins, but the mouse protein is somewhat larger and has a rather different protein sequence.

See also:

Kryokuv G.V. et al. 2003. Characterization of mammalian selenoproteomes. *Science* **300:** 1439–1443.

Which Amino Acids Are Introduced into Proteins by Post-translational Modifications?

In addition to the 20 standard amino acids and the rare selenocysteines that are incorporated into proteins during translation, a large number of post-translationally modified amino acids are present in proteins. Modified amino acids may result from damage reactions but many are introduced by other enzymes encoded in the genome. The table below lists some of these modifications and enzymes that are responsible for their modification. Note the related modifications that may occur, such as hydroxylation of proline at the 3 or 4 positions or multiple methylation of lysine and arginine residues.

Amino Acid	Post-translationally Modified Amino Acid	Proteins in the Modification Pathway
Arginine	Citrulline[1]	PADI family
	Methylarginine[2]	PRMT family
Asparagine	Hydroxyasparagine[3]	ASPH, HIF1AN
Aspartate	Hydroxyaspartate[3]	ASPH
Cysteine	Formylglycine[4]	SUMF1, note also SUMF2 (see reference 4)
Glutamate	Carboxyglutamate[5]	GGCX
Glutamine	Methylglutamine[6]	HEMK1, HEMK2
Histidine	Diphthamide[7]	DPH1, DPH2, DPH3, DPH4, DPH5
Lysine	Hydroxylysine[8]	PLOD family
	Hypusine[9]	DHPS, DOHH
	Methyllysine[10]	DOT1L, SET domain proteins
Proline	Hydroxyproline[11]	EGLN, P4H, LEPRE1 families
Tyrosine	Tyrosine sulfate[12]	TPST1, TPST2

The footnotes are included in "Date Sources, Methods, and References."

The table does not include specialized modifications that occur to the terminal residues of proteins, poly-ADP ribosylation, cross-linking, or histone acetylation. Reactions such as acylation and prenylation are not listed. Likewise, proteolytic processing events and glycosylation reactions are beyond the scope of this work. Protein kinases are discussed in the chapter on gene families (see p. 103).

Data Sources, Methods, and References

The following are the footnotes to the table above:

[1]The peptidylarginine deiminases (PADs) are closely related to each other. See also:

Chavanas S. et al. 2004. Comparative analysis of the mouse and human peptidylarginine deiminase gene clusters reveals highly conserved non-coding segments and a new human gene, *PADI6*. *Gene* **330:** 19–27.

[2]Lee J. et al. 2005. PRMT8, a new membrane-bound tissue-specific member of the protein arginine methyltransferase family. *J. Biol. Chem.* **280:** 32890–32896.

[3]Lando D. et al. 2002. Asparagine hydroxylation of the HIF transactivation domain: a hypoxic switch. *Science* **295:** 858–881.

ASPH is involved in the hydroxylation of both asparagine and aspartate. See:

Korioth F. et al. 1994. Cloning and characterization of the human gene encoding aspartyl beta-hydroxylase. *Gene* **150:** 395–399.

[4]Mariappan M. et al. 2005. Expression, localization, structural, and functional characterization of pFGE, the paralog of the Calpha-formylglycine-generating enzyme. *J. Biol. Chem.* **280:** 15173–15179.

[5]Wu S.M. et al. 1991. Cloning and expression of the cDNA for human gamma-glutamyl carboxylase. *Science* **254:** 1634–1636.

[6]HEMK1 and HEMK2 are related to the products of the yeast genes *MTQ1* and *MTQ2*.

[7]Most of the diphthamide pathway proteins can be identified from their yeast counterparts with BLAST. See also:
Liu S. et al. 2004. Identification of the proteins required for biosynthesis of diphthamide, the target of bacterial ADP-ribosylating toxins on translation elongation factor 2. *Mol. Cell. Biol.* **24:** 9487–9497.

[8]Passoja K. et al. 1998. Cloning and characterization of a third human lysyl hydroxylase isoform. *Proc. Natl. Acad. Sci.* **95:** 10482–10486.

[9]Both of the hypusine pathway enzymes are closely related to their yeast counterparts, DYS1 and LIA1.

[10]Feng Q. et al. 2002. Methylation of H3-lysine 79 is mediated by a new family of HMTases without a SET domain. *Curr. Biol.* **12:** 1052–1058.

[11]Proline hydroxylation occurs at the 3 and 4 positions. See:

Vranka J.A. et al. 2004. Prolyl 3-hydroxylase 1, enzyme characterization and identification of a novel family of enzymes. *J. Biol. Chem.* **279:** 23615–23621.

Epstein A.C. et al. 2001. *C. elegans* EGL-9 and mammalian homologs define a family of dioxygenases that regulate HIF by prolyl hydroxylation. *Cell* **107:** 43–54.

Kukkola L. et al. 2003. Identification and characterization of a third human, rat, and mouse collagen prolyl 4-hydroxylase isoenzyme. *J. Biol. Chem.* **278:** 47685–47693.

Note also the distantly related *PH-4* gene.

[12]Mishiro E. et al. 2006. Differential enzymatic characteristics and tissue-specific expression of human TPST-1 and TPST-2. *J. Biochem.* **140:** 731–737.

Which Sequences Are Associated with the Formylglycine Modification?

Sulfatases have an unusual post-translational modification in which a conserved cysteine is converted to a formylglycine (see p. 93). The human genome encodes 17 genes in the sulfatase family. The figure below shows the pattern of amino acid usage for a 21-aa region centered on the modification site (darker rectangles indicate greater usage).

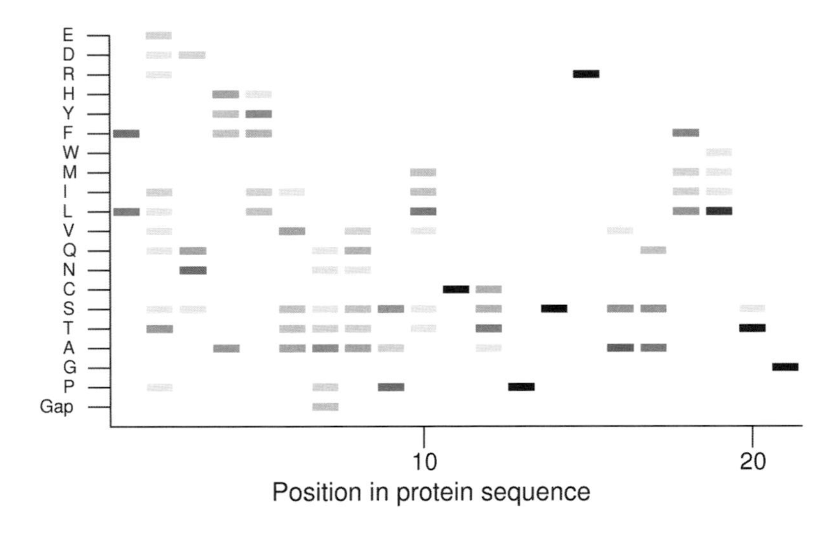

On the carboxy-terminal side of the modified cysteine, there are several completely conserved positions. In contrast, there is comparatively reduced conservation on the amino terminal side of the modification site.

Data Sources, Methods, and References

The protein sequences used to generate the figure were IDS, STS, ARSD, ARSE, GALNS, ARSH, ARSA, ARSF, SGSH (sulfamidase), ARSB, ARSJ, ARSI, GNS, ARSK, SULF1, SULF2, and ARSG. They were readily identified by sequence similarity. STS was used as a query with NCBI BLASTP to generate the alignments, which were then selected manually. Three of the proteins (ARSB, ARSJ, and ARSI) aligned with a gap as shown. The coordinates are for the aligned region, not the intact protein sequences. The protein sequence of a smaller predicted gene (*LOC727913*) was not used for the figure.

Which Proteins Contain γ-Carboxyglutamate?

γ-Carboxyglutamate is a modified amino acid residue found in several components of the coagulation pathway, as well as in a number of other proteins. This modification is performed by GGCX (see p. 93) and requires activated vitamin K. The table below lists proteins that contain γ-carboxyglutamate.

Gene	Protein Product
F2	coagulation factor 2
F7	coagulation factor 7
F9	coagulation factor 9
F10	coagulation factor 10
PROC	protein C (coagulation pathway)
PROS1	protein S (coagulation pathway)
PROZ	protein Z (coagulation pathway)
PRRG1	
PRRG2	
PRRG3	
PRRG4	
GAS6	growth arrest protein 6
BGLAP	osteocalcin
MGP	bone matrix glycoprotein

Data Sources, Methods, and References

Except for BGLAP and MGP, these proteins can be identified by their considerable sequence similarity. PRRG1 (GI:4506135) is an effective query because it does not contain other repeated elements that are found in proteins of the coagulation pathway (see, e.g., coagulation factor II [*F2*], GI:4503635). For more information on BGLAP and MGP, see:

Price P.A. 1989. Gla-containing proteins of bone. *Connect. Tissue Res.* **21:** 51–57.

CHAPTER SEVEN

GENE FAMILIES

A LARGE FRACTION OF HUMAN PROTEINS ARE ENCODED by members of gene families. In this chapter, four broad topics are addressed using intensively studied examples.

The first question deals with the sequence similarity of genes with related functions. There are some families whose members have related functions, but the genes are difficult to assemble based on sequence similarity.

A second issue is the organization of the genes in the family. For recently expanded gene families, the members are often clustered at one location. Members of ancient families often are thoroughly dispersed throughout the genome. It is not unusual to find families that are intermediate between these extremes. Sometimes, linked members are not the most closely related at the sequence level.

For some families a noteworthy characteristic is their extremely large size. With these families, determining their exact size can be challenging because of problems discriminating pseudogenes from functional family members.

Finally, some families are defined more by a small domain or sequence motif than by their overall similarity in sequence. Within these domains or motifs, certain residues are more central for function than others. In these cases, the more diverged members of the larger families can be quite helpful in identifying these residues. Some families are named from certain shared residues. However, other parts of the motif may also be highly conserved.

How Are the DNA Polymerases Related to Each Other?

The human genome encodes many DNA polymerases with diverse roles in the cell. They fall into several families based on sequence similarity. These relationships are summarized in the table below.

Polymerase	Genes	Gene Family Information
α	*POLA, POLA2*	*POLA, POLD1,* and *REV3L* are related
δ	*POLD1, POLD2, POLD3, POLD4*	
ζ	*REV3L*	
ε	*POLE, POLE2, POLE3, POLE4*	*POLE* and *POLE2* each have weak similarity to subunits of the α polymerase
β	*POLB*	all three are related
λ	*POLL*	
μ	*POLM*	
γ	*POLG, POLG2*	unrelated to the others
η	*POLH*	all three are related
ι	*POLI*	
κ	*POLK*	
θ	*POLQ*	*POLN* is related to *POLQ*
ν	*POLN*	

DNTT, which encodes terminal transferase, is related to *POLB*. *REV1L* is related to *POLH*.

POLQ has a domain that is similar to a variety of helicases, notably *HEL308*.

Data Sources, Methods, and References

All genes mentioned above were present in the NCBI RefSeq set that was available when Build 36.2 of the genome sequence was released. Sequence relationships were determined with BLAST 2.2.11 from NCBI.

POLS has been described as polymerase σ, but it is related to yeast *PAP2* and human *PAPD5*.

How Are the Histone Gene Families Organized?

The histone gene families have a complex distribution on the chromosomes. The majority of the histone genes are in a single cluster on chromosome 6 ("Cluster 1"). A second, smaller cluster is on chromosome 1 ("Cluster 2"). In addition, there are unlinked histone genes that are nearly identical to those in the main clusters. Also present in the genome are a variety of genes that encode histone-related proteins. The following table summarizes these relationships. The counts do not include pseudogenes or predicted genes (see also details below).

Histone Type	No. of Genes in Cluster 1	No. of Genes in Cluster 2	No. of Unlinked Genes*	Related Genes
H1	6	0	0	*H1F0, H1FNT, H1FX, H1OO, H1LS1*
H2A	12	4	1	*H2AFB1, H2AFB2, H2AFB3, H2AFJ, H2AFV, H2AFX, H2AFY, H2AFY2, H2AFZ*
H2B	15	2	2	*H2BFM, H2BWT*
H3	10	2	3	*CENPA*
H4	12	2	1	

*Included in the unlinked column are four genes on chromosome 1, located far from Cluster 2.

Cluster 1 includes all five histone types. Some genes in Cluster 1 have significant sequence differences from others in the same cluster. These include *HIST1H1T* (histone 1 family), *HIST1H2BA* (histone 2B family), and *HIST1H4G* (histone 4 family). No annotated histone 1 genes are found in Cluster 2.

The histone-related genes vary in size and in sequence from the prototypes. Included in this category is *CENPA* (centromere protein A), which encodes an H3-related protein. Some of these related genes are clustered.

Data Sources, Methods, and References

The table in this section combines data from RefSeq and the human genome Map Viewer table for release 36.2. After the data were gathered from these sources, *HIST2H3PS2* was labeled as a pseudogene and is not included in the table.

The following predicted genes may also be members of histone gene families: *LOC388177* and *LOC626584* are in the H2A family; *LOC442461* is in the H2B family; and *LOC440093, LOC653604, LOC347376, LOC340096, LOC391769, LOC730740, LOC651863, LOC644914*, and *LOC642186* are in the H3 family.

How Are the Keratin Genes Organized?

The keratins are encoded by a very large gene family. Family members are located in two large clusters on chromosomes 12 and 17, as shown in the figure below. The *KRT-* prefixes have been removed from the names. A region of 1 Mb is shown for each chromosome.

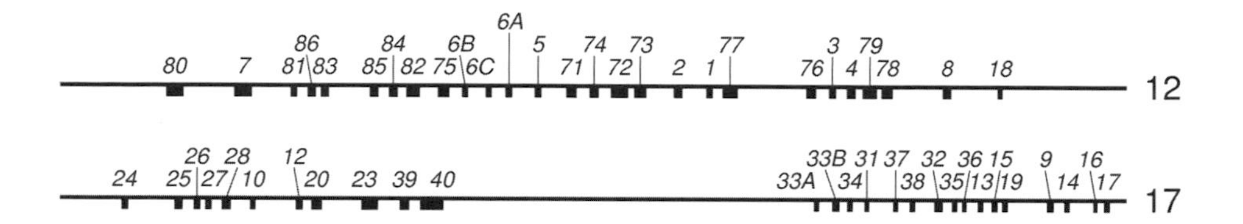

Keratin proteins fall into two classes: The generally more acidic type I chains pair with the relatively more basic type II chains. The gene cluster on chromosome 17 encodes type I keratins. The chromosome 12 cluster encodes type II keratins, with the exception of *KRT18*. *KRT18* encodes a type I chain that pairs with the product of its type II neighbor, *KRT8*.

Within each cluster is a group of genes that encode components of "hard" (generally hair and nail) keratins. The gap in the cluster on chromosome 17 contains many genes for keratin-associated proteins.

Data Sources, Methods, and References

The genes and their coordinates are from release 36.2 of the reference human genome sequence, with the following exceptions: All pseudogenes have been excluded, two annotated genes that have recently been described as pseudogenes, *KRT126P* (type II) and *KRT222P* (type I), have been excluded, and several gene names have been updated (*KRT79* in the figure was *KRT6L*, *KRT6C* was *KRT6B*, and a different gene named *KRT6C* mapped to the same location as *KRT6A*).

See also:

Waseem A. et al. 1990. Embryonic simple epithelial keratins 8 and 18: Chromosomal location emphasizes difference from other keratin pairs. *New Biol.* **2:** 464–478.

How Many Protein Kinases Are Encoded in the Genome?

Most protein kinases are encoded by a large gene family that includes serine/threonine, tyrosine, and dual-specificity kinases. These can generally be identified and placed in subgroups via sequence similarity. There are a smaller number of genes encoding what are classified as atypical protein kinases. Attempts to determine the exact number of protein kinases in the human genome are complicated by the presence of a large number of pseudogenes.

One comprehensive study (Manning et al. 2002) described 478 members of the main kinase family and 40 atypical kinases. This work was extended (Milanesi et al. 2005). The numbers of kinase genes (typical plus atypical) and pseudogenes identified by these investigators are summarized in the table below. Note that both of these studies predate the reference genome sequence release used elsewhere in this volume.

Study	Genes	Pseudogenes
Manning et al. 2002	518	106
Milanesi et al. 2005	523	107

The numbers reported in the table above will likely remain in flux, as various genes and pseudogenes continue to be re-classified. Atypical protein kinases are not readily identified by sequence similarity. In addition, some genes in the main kinase family have differences at otherwise highly conserved residues. Additional discovery of polymorphisms may also lead to revised interpretation of particular loci.

Data Sources, Methods, and References

Manning G. et al. 2002. The protein kinase complement of the human genome. *Science* **298:** 1912–1934.

Milanesi L. et al. 2005. Systematic analysis of human kinase genes: A large number of genes and alternative splicing events result in functional and structural diversity. *BMC Bioinformatics* (suppl. 4) **6:** S20.

Which Types of Proteases Are Found in the Human Genome?

Unlike the protein kinases, where most of the genes can be placed in one large family through sequence similarity, there are a number of classes of proteases present in the genome with comparable representation. Pseudogenes, mutations in key residues, and the presence of unrelated types all present challenges to counting proteases. Continued analysis has resulted in small changes to the totals.

The table below presents the number of human protease genes in each of the mechanistic classes.

Class	No. of Genes
Aspartate proteases	21
Cysteine proteases	148
Metalloproteinases	186
Serine proteases	178
Threonine proteases	28
Total	561

Data Sources, Methods, and References

The numbers in the table are from:

Puente X.S. and López-Otín C. 2004. A genomic analysis of rat proteases and protease inhibitors. *Genome Res.* **14:** 609–622.

Comparative genomics has had a major role in accounting for protease family members. See:

Puente X.S. et al. 2005. Comparative genomic analysis of human and chimpanzee proteases. *Genomics* **86:** 638–647.

How Many Genes Are in the Major Transcription Factor Families?

Genes for human transcription factors are often found in large families. The table below includes the approximate number of genes in several transcription factor families, as well as some examples of well-known members.

Family	No. of Genes	Examples
Krüppel-type	>650	KLF family
Homeobox	~190	HOX genes
Helix-loop-helix	~110	MYC
Nuclear receptors	47	AR
FOX	46	Hepatocyte nuclear factor 3 family (FOXA1 and related)
ETS	28	ETS1 and related
SOX	20	SRY
T-box	17	T
POU	15	Octamer-binding proteins (POU2F1 and related)

The largest family is, by far, the Krüppel-type zinc finger proteins. In this unusually large group, functional information exists for only a fraction of the family members. Some of the genes may have other functions.

Many of the families presented in the table are readily gathered by sequence similarity starting with well-known prototypical members. Some of the family members are quite different in size from the others in their group, but the domain leading to the assignment is generally clear. Some distant branches of the homeobox family are less readily defined. The helix-loop-helix family is also a diverse group. As described on page 106, the POU family is related to the homeobox proteins.

Data Sources, Methods, and References

The counts above are from the loci included in human RefSeq. Relationships were determined by searches with BLASTP. Some predicted genes and more distant family members have been excluded from the counts. Literature estimates differ from the table above for a variety of reasons, including their source set of sequences and the criteria for inclusion in the counts. For example, some estimates place almost 800 genes in the Krüppel-type zinc finger family.

For additional information on transcription factor families in the genome, see:

Lander E.S. et al. 2001. Initial sequencing and analysis of the human genome. *Nature* **409:** 860–921.

Messina D.N. et al. 2004. An ORFeome-based analysis of human transcription factor genes and the construction of a microarray to interrogate their expression. *Genome Res.* **14:** 2041–2047.

Venter J.C. et al. 2001. The sequence of the human genome. *Science* **291:** 1304–1351.

For other estimates of the size of the Krüppel-type zinc finger family, see:

Looman C. et al. 2002. KRAB zinc finger proteins: An analysis of the molecular mechanisms governing their increase in numbers and complexity during evolution. *Mol. Biol Evol.* **19:** 2118–2130.

Rousseau-Merck M.F. et al. 2002. The KOX zinc finger genes: Genome-wide mapping of 368 ZNF PAC clones with zinc finger gene clusters predominantly in 23 chromosomal loci are confirmed by human sequences annotated in EnsEMBL. *Cytogenet. Genome Res.* **98:** 147–153.

Which Are the Important Residues in the Homeobox?

The homeobox transcription factor family includes about 190 genes (see the notes at the end of this section) with varying levels of sequence conservation. While these genes vary considerably in size, generally, they can easily be identified by the conserved homeobox domain.

In the figure below, 28 family members were assembled to show the diversity of amino acids found at various positions in the homeobox domain. Darker boxes indicate higher levels of amino acid usage at a given position (a black box indicates complete conservation across the selected set of proteins). Some homeobox family members align over longer regions than the 57-aa segment shown in the figure.

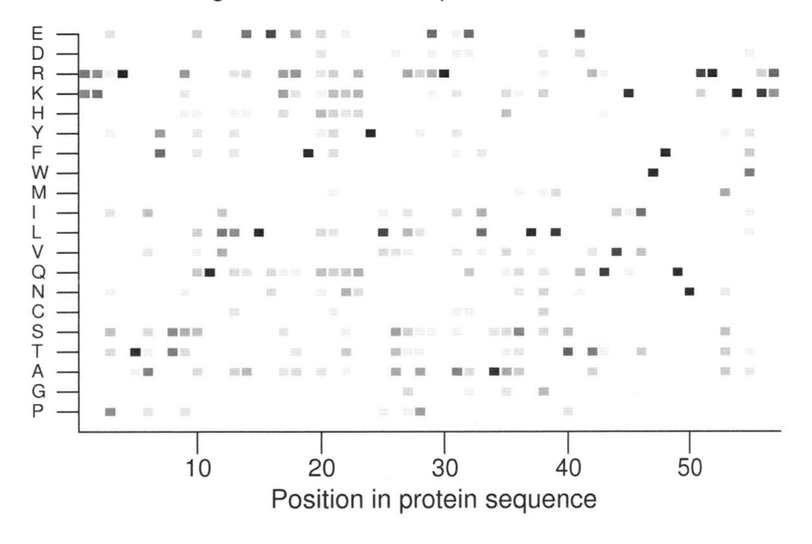

As shown in the figure, many positions have conserved basic amino acids, including those at the ends. In this set, there are completely conserved glutamine, phenylalanine, and tyrosine residues at positions 11, 19, and 24, respectively. Another conserved segment is the sequence of tryptophan, phenylalanine, glutamine, and asparagine, which spans positions 47 to 50 and is present in more than two-thirds of the entire family. In one small branch of the family (*BARX1* and related genes), the phenylalanine at position 48 is a tyrosine. The most common variation at position 49 is a change from glutamine to lysine (this occurs in the pituitary [PITX] and sine oculis [SIX] homeobox types).

The POU family is a diverged group of homeobox proteins. All members of this family, except for a single predicted gene, have a conserved cysteine instead of a glutamine at position 49 of the homeodomain. As shown in the following figure, some of the other conserved residues in the more common homeobox types are also different in the POU proteins.

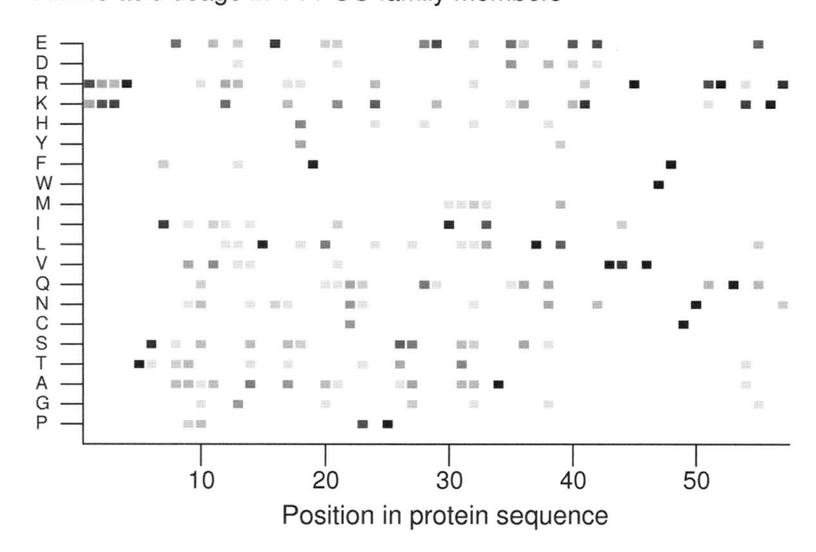

Data Sources, Methods, and References

The estimate of 190 genes includes the 39 HOX genes and many more diverged family members, but not the related POU family, which has 15 genes (see p. 105). The proteins used in the first figure were BAPX1, DLX1, DLX2, DLX3, DLX4, DLX5, DLX6, EMX1, EMX2, EVX1, GBX2, HOXC4, HOXB4, HOXA4, HOXD4, HLXB9, HMX1, HMX2, IPF1, LBX1, MEOX1, MEOX2, MSX1, MSX2, RAX, TLX1, TLX2, and TLX3. The aligned regions were assembled using NCBI BLAST 2.2.11 and the query sequence DLX1. The selected proteins produced ungapped alignments for the 57-aa region shown.

For the second figure, all of the POU family members except POU5F2 (FLJ25680) were used. Two of the proteins did not align with the query POU1F1 at the final position using NCBI BLAST 2.2.11, and those were added manually.

Coordinates in both figures relate to the sequence alignments, not the complete proteins.

How Are the HOX Genes Organized?

The HOX genes are a major subclass of the homeobox genes. They are found in four clusters on chromosomes 2, 7, 12, and 17, as shown in the figure below (a 200-kb segment for each chromosome is drawn to scale; the *HOX*- prefix on each gene has been removed).

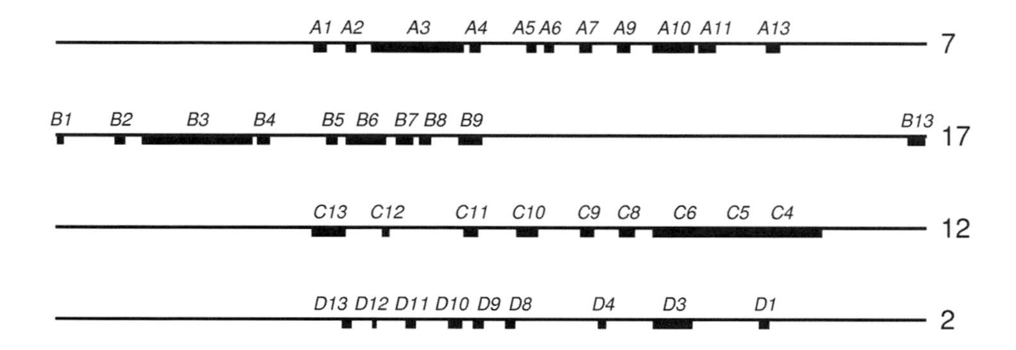

The 39 HOX genes are divided into 13 types. Various combinations of genes are present in the four clusters, but the order and orientation of the genes within the clusters is conserved. The direction of transcription is from the higher-numbered genes toward the lower-numbered genes (right to left in the A and B clusters; left to right in the C and D clusters). No cluster has all 13 types and 11 types are found in the A cluster.

The B cluster is considerably larger than the others, but only has 10 of the types represented. Note the large gap between *HOXB9* and *HOXB13*. One known gene, *PRAC*, is located in this interval close to *HOXB13*. Two other predicted genes have been reported in that region.

The *HOXC6*, *HOXC5*, and *HOXC4* genes are shown as a single block because they have been found, in some cases, to share a common 5′ noncoding exon. Although the protein products are similarly sized, the genes in the clusters vary considerably in the spans of their transcribed regions.

Two small predicted genes have also been annotated in the D cluster (not shown). Additional small transcripts other than those mentioned here have been annotated in these regions.

Data Sources, Methods, and References

The data used to produce the figure are from the NCBI Map Viewer. Genes flanking the clusters are not shown. The orientations are those on the reference chromosome sequences.

See also:

Simeone A. et al. 1988. At least three human homeoboxes on chromosome 12 belong to the same transcription unit. *Nucleic Acids Res.* **16:** 5379–5390.

Which Sequences Are Shared in Helix-loop-helix Transcription Factors?

Another large group of transcription factors is the helix-loop-helix type. More than 100 examples are found in the human genome. Among these, one large subgroup consists of 24 genes related to *Drosophila atonal* and *scute*, the neurogenin family, and others identified through rearrangements in T-cell tumors. The products of these genes can be aligned over a 49-aa region without gaps. The pattern of their sequence conservation is presented in the figure below, with dark boxes indicating a higher proportion of amino acid usage at a given position.

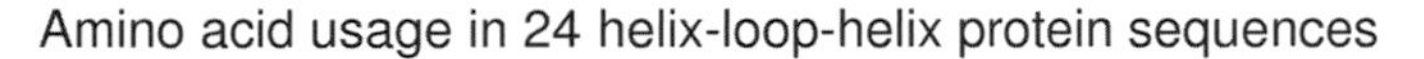

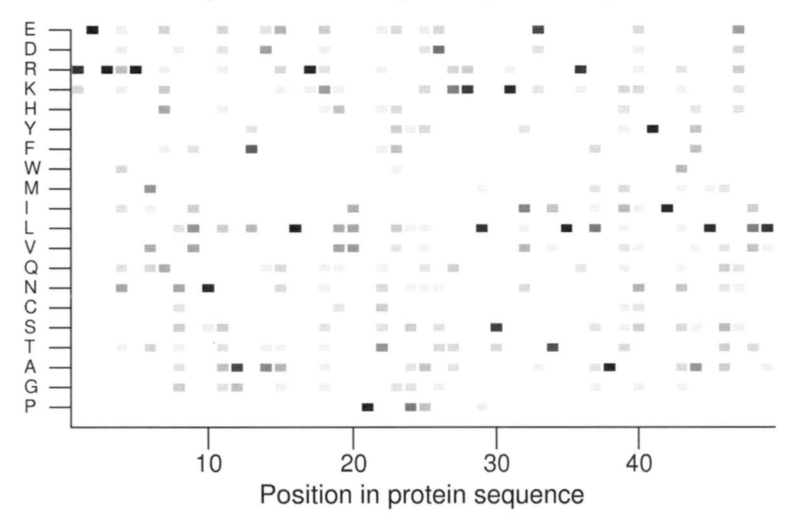

In many cases, these alignments can be extended by additional positions, particularly amino-terminal to the region shown. Other subgroups of this family align with gaps (not shown).

Data Sources, Methods, and References

The sequences used to generate the figure were found using amino acids 122–175 of the protein encoded by the *ASCL1* gene as a query in a search with BLASTP. Sequences of the following proteins were used: ASCL2, ASCL3, ASCL4, ATOH1, ATOH7, ATOH8, FERD3L, FIGLA, HAND1, HAND2, LYL1, NHLH1, NHLH2,

NEUROD1, NEUROD2, NEUROD6, NEUROG1, NEUROG2, NEUROG3, PTF1A, TAL1, TAL2, and TCF15. These sequences were manually trimmed to align with positions 126–174 from ASCL1.

Which Residues Define the DEXD Helicases?

The DEXD helicase family includes about three dozen genes. Although the family is named for its DEXD tetrapeptide motif, other residues are highly conserved. In the figure below, 17 DEXD family members were aligned along a 35-aa segment to identify other conserved residues. Darker boxes in the figure indicate higher conservation (black boxes indicate complete identity among the 17 proteins).

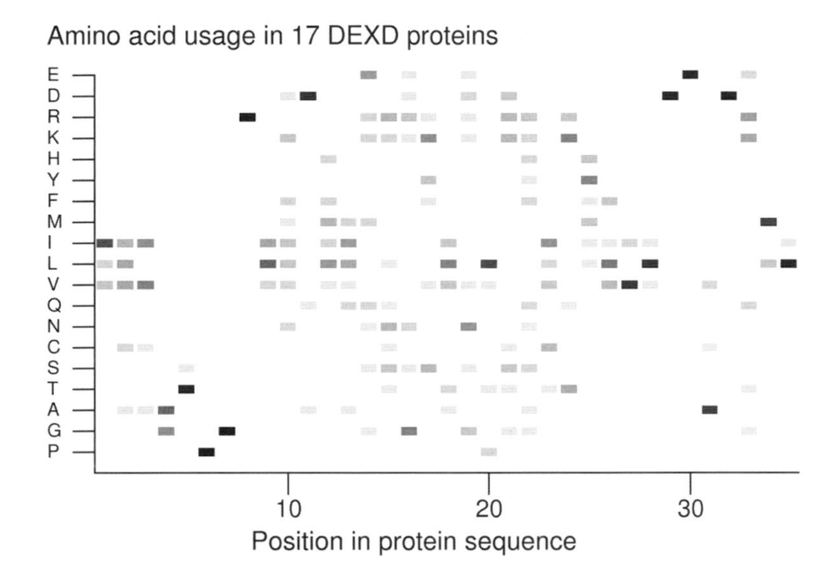

The DEXD motif is readily seen at positions 29 to 32. Most of the proteins have an A at the variable position (X), giving rise to the "DEAD box" name that is often used. The figure shows a second motif, PGR, at positions 6–8. In this set of proteins, the PGR motif is completely conserved in sequence and in spacing relative to the DEXD motif.

Data Sources, Methods, and References

The proteins used for the figure were DDX1, DDX3X, DDX3Y, DDX4, DDX5, DDX6, DDX17, DDX20, DDX21, DDX23, DDX43, DDX48, DDX50, DDX53, DDX59, BAT1, and EIF4A1. These proteins produced ungapped alignments using NCBI BLAST 2.2.11. The coordinates are from the aligned region, not the intact sequences.

What Are the Conserved Features of the WD Domain?

The WD domain is readily identified in about 200 human proteins. Many of these proteins have multiple copies (repeats) of the WD domain. It is named for the WD dipeptide, which is relatively rare in proteins but generally defines the carboxy-terminal end of the WD domain. In many family members, a conserved GH dipeptide is located amino-terminal to the WD sequence.

Alignments of WD repeats often contain gaps. To create the figure below, ten proteins that yielded ungapped alignments were used (darker boxes indicate higher levels of conservation). In each of these ten proteins, the GH dipeptide is separated by 27 amino acids from the WD dipeptide.

Note the limited diversity around position 20 and the conserved aspartate at position 24. Otherwise, even with this relatively conserved set, there is substantial variation.

The WD domain is sometimes defined as having additional residues amino-terminal to the region shown. These positions also have a high level of variation.

Data Sources, Methods, and References

One repeat unit from each of the following proteins was used to produce the figure: GNB1 (182–212), WDR3 (189–219), WDR69 (175–205), GNB2L1 (190–220), TAF5L (508–538), NLE1 (412–442), WDR36 (620–650), TAF5 (667–697), WDR61 (230–260), and APG16 (322–352). The alignments were generated with NCBI BLAST 2.2.11.

How Are the Olfactory Receptor Genes and Pseudogenes Distributed in the Genome?

A total of 370 genes and 467 pseudogenes from the major olfactory receptor family have been annotated on the chromosomes. This family is, by far, the largest class of G-coupled receptors. The table below shows the chromosomal assignments for the genes and pseudogenes in this family.

Chr.	No. of Genes	No. of Ψ	Chr.	No. of Genes	No. of Ψ
1	57	35	13	0	8
2	2	8	14	21	14
3	10	21	15	5	13
4	0	11	16	2	1
5	3	5	17	13	4
6	15	21	18	0	2
7	15	20	19	19	22
8	1	10	20	0	0
9	25	14	21	0	3
10	1	7	22	1	0
11	163	203	X	1	10
12	16	25	Y	0	0

On a given chromosome, the genes may be in multiple locations (including clusters). There are many isolated genes and pseudogenes in the olfactory receptor family.

As shown in the table, almost half of the family is on chromosome 11. Smaller clusters are found on other chromosomes.

Five human genes (*VN1R1, VN1R2, VN1R4, VN1R5,* and *FKSG83*) are related to vomeronasal receptors of the mouse. Their function is not clear, but they have some similarity to members of the type 2 taste receptor family (for a discussion of the taste receptor family, see p. 116). Three annotated, unlinked pseudogenes are also present. Three of the genes (*VN1R1, VN1R2,* and *VN1R4*) are located on chromosome 19.

Data Sources, Methods, and References

Discrimination betweeen genes and pseudogenes in the olfactory receptor family can be difficult. The totals above are from the chromosome annotations in release 36.2 of the genome sequence. The RefSeq set available at that time included 376 olfactory receptor genes.

How Are the Taste Receptor Genes and Pseudogenes Distributed in the Genome?

The taste receptor families are much smaller than the olfactory receptor families. There are three genes for the type 1 taste receptors. They are related to a variety of other G-coupled receptors. All three type 1 genes are located on chromosome 1 but are not tightly linked to one another.

Most taste receptors are in the type 2 family, which is a more distant branch of the G-coupled receptor family. The table below gives the chromosomal assignments for the type 2 taste receptor genes and pseudogenes.

Chr.	No. of Genes	No. of Ψ
5	1	0
7	9	2
12	14	5

About half of the type 2 taste receptor genes are in a single cluster on chromosome 12. Isolated genes are present on all three chromosomes.

PKD1L3 and *PKD2L1* may function in sensing sour taste but are not included in the table.

Data Sources, Methods, and References

The totals above are from the chromosome annotations in release 36.2 of the genome sequence.

The RefSeq set for the type 2 taste receptors has one additional gene, *TAS2R45*. *LOC728338* is *TAS2R47* and is included in the table.

See also:

Ishimaru Y. et al. 2006. Transient receptor potential family members PKD1L3 and PKD2L1 form a candidate sour taste receptor. *Proc. Natl. Acad. Sci.* **103:** 12569–12574.

CHAPTER EIGHT

MOBILE ELEMENTS AND REARRANGING GENES

ALMOST HALF OF THE HUMAN GENOME CAN BE READILY ASSIGNED to various classes of highly repeated sequences. Most of these repeated sequences are the well-known types of transposable elements. Transposable elements can be divided into two broad classes: those that use an RNA intermediate during replication and those that do not. The human genome contains representatives of both classes that fall into several families.

A major feature of the main families of transposons is the high fraction of fragmentary or otherwise defective copies of these elements. Both the degree of fragmentation and sequence divergence can be used to provide clues as to when these elements entered the genome. Additional evidence can be obtained via comparative studies of genomes. Elements that are currently active, as well as those that have moved more recently, remain largely intact and have diverged less from their consensus sequences.

Many of the human families of mobile elements are not capable of autonomous movement and rely on transposition functions provided by others. Some of the transposon families, especially those that lack an RNA intermediate, appear to be completely inactive in humans.

Also presented in this chapter is information about the genes that undergo rearrangement during the development of the immune system. Although these genes occupy a small fraction of the overall genome, their role in defense has led to detailed analysis of their mechanism of rearrangement.

How Much of the Genome Is Composed of Mobile Elements?

Approximately 45% of human genome sequences are derived from the major classes of transposable elements that are listed in the table below.

Class		Genome Fraction (%)
Transposons with RNA intermediates	SINE	13
	LINE	21
	LTR	8
DNA transposons		3

Among the nonautonomous SINE elements, Alu types form the largest fraction, covering more than 10% of the human genome. The L1 types of autonomous LINE elements cover more than 17% of the genome. Additional information about the various transposon classes is found in subsequent sections of this chapter.

Besides mobile elements, additional repeats are present in the sequenced fraction of the genome. Almost 1.5% of the sequenced fraction of the genome is a variety of low-complexity sequences. Another source of repeats is processed pseudogenes.

Data Sources, Methods, and References

The numbers above are based on the NCBI 36.2 genome assembly. They may differ slightly from literature values that are based on earlier assemblies.

See also:

Deininger P.L. and Batzer M.A. 2002. Mammalian retroelements. *Genome Res.* **12:** 1455–1465.

Lander E.S. et al. 2001. Initial sequencing and analysis of the human genome. *Nature* **409:** 860–921.

Which Types of L1 Elements Are Present in the Genome?

The L1 family is the most abundant of the LINE-type elements. Approximately 900,000 L1-related regions have been annotated onto the chromosomes. When adjacent or overlapping L1 annotations are merged (see the details at the end of this section), this total is reduced to about 800,000 segments. Most L1 sequences fall into two subtypes based on their taxonomic distribution: mammalian (about 661,000) and primate (about 154,000). A third subtype, the human L1 elements, are much less numerous (a little over 1000).

The figure below shows the size distribution of these three subtypes. For each subtype, the plot shows the cumulative number of segments that are smaller than the indicated size (50% on the *y*-axis indicates the median).

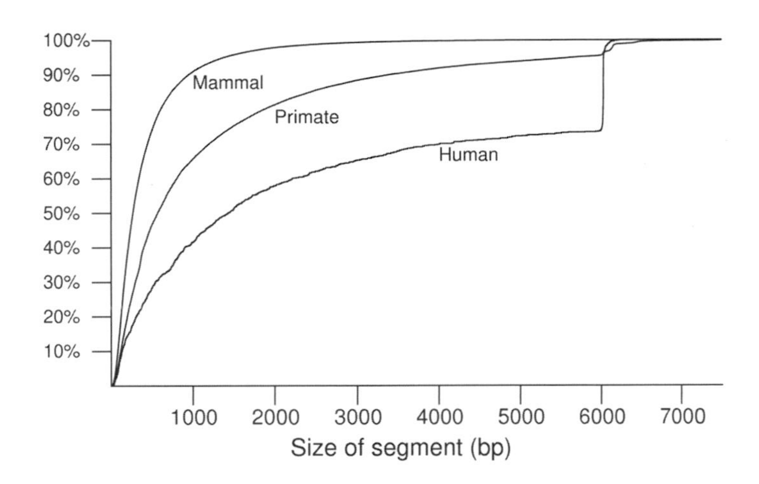

Most L1 sequences are relatively small fragments that have been generated by incomplete reverse transcription or by rearrangements of the genome. The latter mechanism can be used to infer the age of transposition events. The older mammalian subtypes are typically smaller than the primate subtypes. Virtually all mammalian subtype segments are smaller than 3 kb. Although most of the primate subtypes are present as small fragments, a significant number are greater than 3 kb, and a small fraction is a little over 6 kb, the size of a complete element. For the L1 human subtypes, about 30% of the elements are near full-length. Segments larger than unit size likely arose by the transposition of segments into existing elements or by other rearrangements that yielded similar structures.

In the following figure, the size and chromosomal distribution of the L1 human subtypes is presented. Each segment is plotted at the position corresponding to its size and chromosomal assignment. Near-full-length elements are present on most of the chromosomes. Few human L1 segments are present on the most gene-rich chromosomes, but some large copies are present.

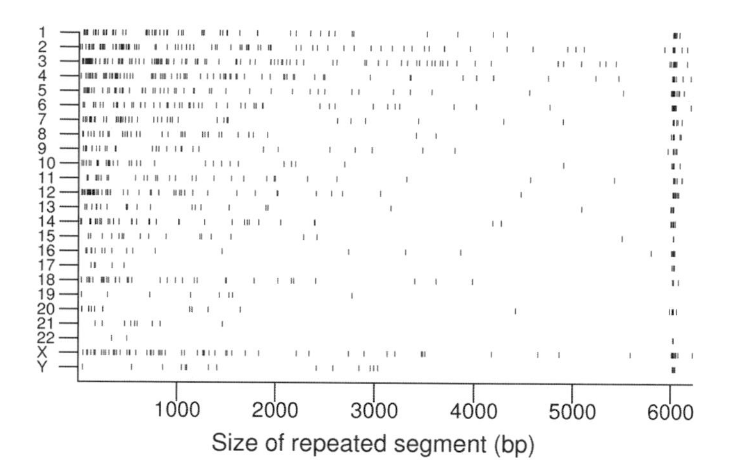

Data Sources, Methods, and References

The figures in this section were generated from the table of repeats annotated onto release 36.2 of the reference genome sequence. All entries with names beginning with L1 were collected. Because of the methods used during the annotation process, adjacent or overlapping segments may have related annotations. In this analysis, such segments were merged, regardless of orientation. Because some classes of transposons have inverted repeats, this approach is helpful in trying to detect larger functional units. As indicated above, this reduced the count of L1-related segments by about 100,000.

The first figure presents the integrals of the histograms for the mammalian, primate, and human L1 elements (after merging the segments with the related annotations L1M, L1P, and L1HS, respectively). For the mammalian and primate subtypes, a few of the merged copies were much larger than unit size. These were used when the counts were normalized, but they are not shown because the plots were truncated at 7500 bp.

For the second figure, there were a total of 1171 human L1 segments, 1089 of which were 100 bp or larger and 685 of which were 1000 bp or larger. These numbers reflect the merge of overlapping segments and related annotations.

See also:

Salem A.H. et al. 2003. LINE-1 preTa elements in the human genome. *J. Mol. Biol.* **326:** 1127–1146.

What Is the Number and Size Distribution of Alu Family Sequences in the Genome?

Alu repeats are even more numerous than the L1 elements described on page 119 in the previous section. More than one million of these very short elements are present. Collectively, they cover about 10% of the genome. Alu elements are not autonomous and rely on the L1 transposition system for movement. Although some fragments of Alu repeats are present in the genome, as shown in the figure below (bin size of 1), most Alu sequences are in a single size class near unit size.

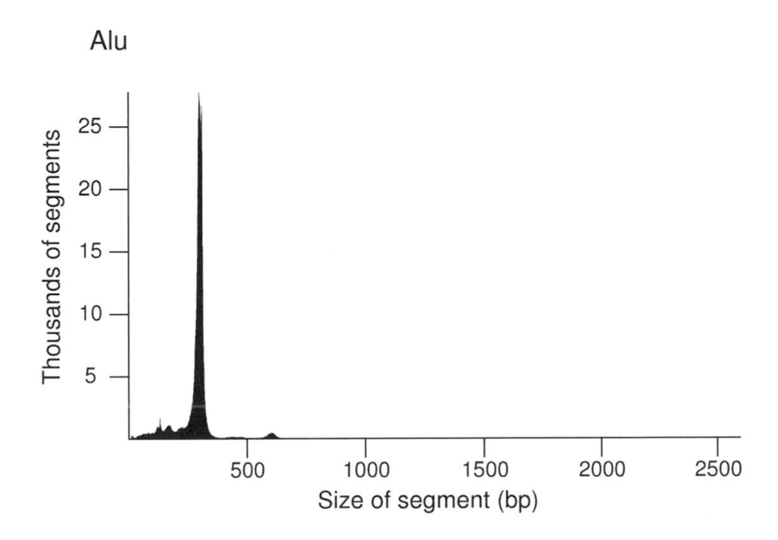

This figure was produced after merging the overlapping and adjacent repeats. The x-axis was scaled to cover all observed sizes. The largest Alu-annotated regions are rare, and they are not visible on the graph because of the scale on the y-axis. These large Alu-related segments generally have a lower complexity component.

In this graph, the mode is at 299 bp. There is a small peak at about 600 bp, which is approximately twice the unit length. These segments may have been derived from insertions of full-length units at existing Alu elements. Because about 10% of the genome is composed of Alu elements, if the insertions were random, one would expect this peak to be larger.

Only a small fraction of Alu elements are in subfamilies associated with recent movement events. One such subfamily, Alu Yb8, contains 638 merged, annotated segments in the genome. The segments from this subfamily show a tight distribution in plots like the one above, with about one-sixth of the repeats being 318 bp.

Data Sources, Methods, and References

The data for this section were obtained and analyzed in the same manner as described for the L1 family on page 120. Sequences with names beginning with Alu were used to prepare the figure. A total of 1,029,378 segments (after the merge of adjacent and overlapping annotations) were found. Other related repeats are present in the genome.

See also:

Carroll M.B. 2001. Large-scale analysis of the Alu Ya5 and Yb8 subfamilies and their contribution to human genomic diversity. *J. Mol. Biol.* **311:** 17–40.

Which Are the Common DNA Transposons in the Genome?

DNA transposons represent about 3% of the sequenced human genome and are not believed to have been active for tens of millions of years. Although there are a very large number of DNA transposon families, most transposons fall into two broad classes. One class, known as hAT, includes the abundant Charlie types. The second major class, which includes transposons such as Tc1 and mariner, is most commonly represented by the Tigger types. The mariner transposons are less numerous than the Tigger transposons.

The Charlie types are believed to have entered the genome at an earlier point in evolution. As shown in the figure below (bin size of 10), the Charlie-related sequences in the genome have, for the most part, been extensively fragmented. When the Charlie subfamilies are examined separately, more structure in the size distribution is seen (not shown).

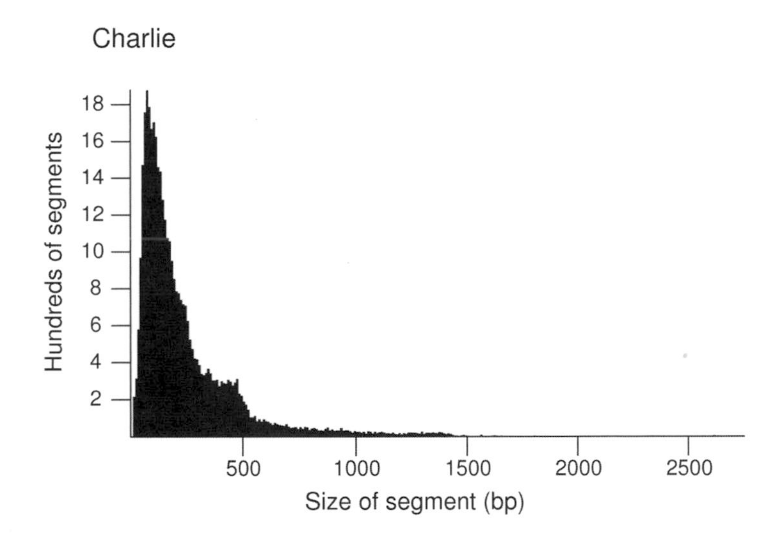

The Tigger types, shown in the figure on the next page (also bin size of 10), are thought to have entered the genome somewhat later in evolution. There is considerably more structure in the size distribution; much of this is derived from differences in the subfamilies. Because they have been in the genome for a shorter period of time, there has not been as much fragmentation of the elements. The peak at about 2400 bp is from the copies of the Tigger1 subfamily that are approximately full-length.

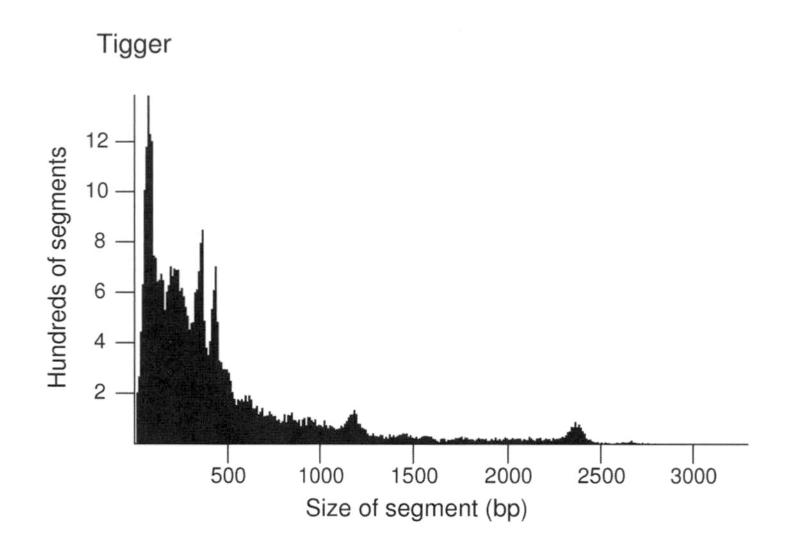

Data Sources, Methods, and References

The figures in this section were generated from annotation 36.2 of the NCBI reference human genome sequence. All regions where annotation began with Charlie (41,288 annotated segments) or Tigger (46,271 annotated segments) were used. In both cases, the annotation classifies these into a number of subfamilies. As described in the section on L1 elements (p. 120), overlapping segments were merged, regardless of orientation. After merging these segments, a total of 39,670 Charlie segments and 41,884 Tigger segments remained. Both axes were autoscaled to include the bin with the largest segment and the bin with the largest count.

See also:

Giordano J. et al. 2007. Evolutionary history of mammalian transposons determined by genome-wide defragmentation. *PLoS Comput. Biol.* **3:** e137.

Lander E.S. et al. 2001. Initial sequencing and analysis of the human genome. *Nature* **409:** 860–921.

Pace J.K. and Feschotte C. 2007. The evolutionary history of human DNA transposons: Evidence for intense activity in the primate lineage. *Genome Res.* **17:** 422–432.

Which Endogenous Retroviral Genomes Are Largely Intact?

Although much of the repeated fraction of the human genome contains LTR and internal retroviral sequences, most of the retroviral genomes have become extensively fragmented and have diverged at the sequence level since their introduction into the human genome. One subgroup of HERV-K elements remains largely intact. The members of this subgroup typically differ by about 2% at the nucleotide level, and many small insertions and deletions are present. The table below lists the locations of the more intact and conserved elements in the reference genome sequence, and includes notes about certain deletions in some of the copies. Several have a common deletion near the env gene.

Nearby Gene	Chr.	Comments
MSTO1	1	deletion near env
LOC25871	3	deletion near env
SENP2	3	deletion near env
CDH6	5	
SGCD	5	deletion near env
HTR1B	6	deletion near polymerase
FOXK1	7	tandem copy
FOXK1	7	tandem copy
DEFB107B	8	K115 sequence, located in intron
TRPC6	11	
LOC338805	12	
PRODH	22	deletion near env

Some of the integration sites are polymorphic in the human population. Note that the table lists the elements in the reference genome sequence, which contains the K115 element on chromosome 8 but lacks the K113 element on chromosome 19 (for more details on these elements, see below).

Data Sources, Methods, and References

The elements in this table were easily identified using the K115 element (GI:16507983) to search the genome with BLASTN. GI:9558700 contains the tandem copies on chromosome 7. More diverged HERV-K copies are found near *LOC439949* (on chromosome 10) and *DDX6* (on chromosome 11; it also contains a large deletion).

For additional information on the K115 and K113 elements, see:

Turner G. et al. 2001. Insertional polymorphisms of full-length endogenous retroviruses in humans. *Curr. Biol.* **11:** 1531–1535.

Are the Immunoglobulin and T-cell Receptor Loci Very Large?

The rearranging genes for immunoglobulin (Ig) and T-cell receptor subunits are at six chromosomal locations. Although these loci are generally large, they do not cover a significant fraction of the genome, and in their unrearranged forms, would not rank among the very largest human genes, which are greater than 2 Mb (see p. 26). The size of each locus (rounded to the nearest 10 kb) is listed in the table below.

Locus	Chr.	Size (Mb)
Ig heavy chain	14	1.23
Ig κ	2	0.97
Ig λ	22	0.88
T cell α/δ	14	0.93
T cell β	7	0.58
T cell γ	7	0.13

The loci for each of these genes contain many variable (V) segments, diversity (D) segments (for some chains), joining (J) segments, and constant gene segments. Each locus also contains a large number of nonfunctional (pseudogene) segments. Unrelated genes may also be present (e.g., members of trypsin family are at the T-cell receptor β locus). Additional information about the internal organization of each locus is described on pages 128–132.

In the figure on the following page, the chromosomal locations of these six genes and some nonrearranging loci with related functions are presented.

The immunoglobulin λ (*IGL*) locus occupies a significant segment of the rather small chromosome 22. The β and γ T-cell receptor loci on chromosome 7 (*TRB* and *TRG*, respectively) are not closely linked. On chromosome 14, the α/δ T-cell receptor locus (*TRA/D*) is not closely linked to the immunoglobulin heavy chain locus (*IGH*). The *IGH* genes are essentially located at the telomere of chromosome 14.

PTCRA, *IGLL1*, *VPREB1*, and *VPREB3* are small genes (less than 10 kb). *PTCRA* functions as an α chain during T-cell development but is not linked to any of the rearranging T-cell receptor loci. In contrast, the *VPREB1* and *IGLL1* genes, which function as light chains during B-cell development, are closely linked to the λ light chain (*IGL*) locus. *VPREB3* is related to *VPREB1*. A number of other related sequences are not shown.

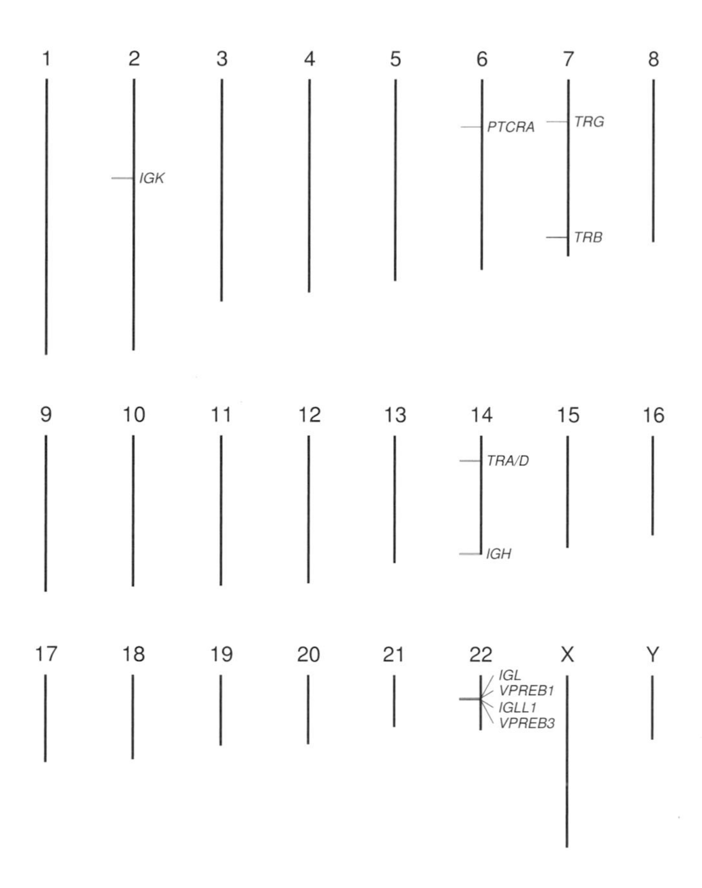

Data Sources, Methods, and References

The locus sizes in the table were based on the coordinates of the gene segments mapped onto the reference human genome sequences in release 36.2. The sizes include all gene segments, regardless of orientation, as well as any unrelated genes present in these chromosomal regions. The size of the Ig κ locus is the span of the two closely linked groups of gene segments. The chromosomal locations in the figure were from the Map Viewer tables.

See also:

Bankovich A.J. et al. 2007. Structural insight into pre-B cell receptor function. *Science* **316:** 291–294.

Del Porto P. et al. 1995. Cloning and comparative analysis of the human pre-T-cell receptor alpha-chain gene. *Proc. Natl. Acad. Sci.* **92:** 12105–12109.

How Are the Immunoglobulin Genes Organized?

Immunoglobulin (Ig) genes are present at three locations in the genome (see p. 127). Different numbers of V (variable) segments, J (joining) segments, and constant region genes are present at each locus. D (diversity) segments are also present at the heavy chain locus. These gene segments can rearrange to encode functional polypeptide chains. The counts of these genes are summarized in the table below (pseudogenes are indicated with the symbol Ψ).

Locus	V Segments	D Segments	J Segments	Constant Genes
Ig heavy chain	$44 + 79\Psi$	27	$6 + 3\Psi$	$9 + 2\Psi$
Ig κ	$41 + 31\Psi$	0	5	1
Ig λ	$38 + 35\Psi$	0	7	$5 + 2\Psi$

The heavy chain locus has another layer of complexity. It can produce different classes of antibodies by switching the segments encoding the constant region. The order of the constant segments (including the pseudogenes) at the heavy chain locus is shown below.

$$\mu - \delta - \gamma 3 - \gamma 1 - \varepsilon\psi 1 - \alpha 1 - \gamma\psi - \gamma 2 - \gamma 4 - \varepsilon - \alpha 2$$

The organization of the κ locus is distinct, with the V genes being present in two large clusters. Both clusters of V genes are 5′ to the constant region gene. Although the V genes in the proximal cluster are generally in the same orientation as the constant region gene, those in the distal cluster are inverted.

At the λ locus, one J segment is located 5′ to each constant region gene (there are seven J segments and seven constant gene segments, including the pseudogenes).

Data Sources, Methods, and References

The data in this table are from the assignments on release 36.2 of the reference genome sequence and include segments identified as pseudogenes. Orphan segments are not included in the totals. Additional immunoglobulin-related pseudogenes and fragments may not be annotated on the reference genome.

How Are the T-cell Receptor Loci Organized?

There are three loci that encode the T-cell receptors (see p. 126). They are moderately large and include many segments that rearrange during development to produce genes that encode functional proteins. The table below summarizes the numbers of gene segments at each locus.

Locus	V Segments	D Segments	J Segments	Constant Genes
T cell α/δ	49 + 8Ψ	3	4	δ
		0	58 + 3Ψ	α
T cell β	51 + 16Ψ	1	6	C1
		1	7 + 1Ψ	C2
T cell γ	6 + 8Ψ	0	3	C1
		0	2	C2

Each locus has two constant region genes. The assignment of the D (diversity) and J (joining) segments in the table is based on their being 5′ to particular constant region genes. D segments are present only for the β and δ chains. Many pseudogenes for the V and J segments have been characterized at these loci, and these are indicated in the table by the symbol Ψ.

At the α/δ locus, there is a complex pattern of usage of particular V segments with the constant region genes. Relatively few V segments are used in δ chains. Two of the V segments at the α/δ locus are beyond unrelated genes at the 5′ end of the region (orientation relative to the constant region genes). Another V segment at that locus is located between the constant region genes and is inverted. Similarly, a V segment is 3′ of the constant genes at the β locus and in an inverted orientation.

Data Sources, Methods, and References

The data in the table are from the assignments on release 36.2 of the reference genome sequence and include those segments identified as genes or pseudogenes. Orphans and other fragments are not included in the counts. The V segment counts are the totals at each locus, regardless of position or orientation. Some of the gene segments for T-cell receptors (including the J γ gene segments) have names suggesting they are pseudogenes but are listed in the Map Viewer tables as genes. Similarly, some pseudogenes are not obvious from their names.

Which Sequences Direct DNA Joining at the Immunoglobulin and T-cell Receptor Loci?

V, D, and J segments join to form mature receptor genes in the immune system. In these reactions, two regions of DNA are involved, one at the end of each gene segment. Each region has two short, conserved sequences that are 7 and 9 nucleotides in length. The two regions differ in spacing between the conserved 7- and 9-nucleotide elements; one region generally has 12 nucleotides between the conserved elements and the other generally has 23 nucleotides between the conserved elements. The configuration of these regions at the Ig λ locus is presented in outline fashion below:

V region CACAGTG ← 23 nucleotides → ACAAAAACC →

← GGTTTTTGT ← 12 nucleotides → CACTGTG J region

Note the complementarity of the conserved 7- and 9-nucleotide sequences. The pairing rules are such that two gene segments with 12-nucleotide spacers (or two gene segments with 23-nucleotide spacers) will not direct joining. The orientation of these regions relative to the sense strand for translation is not conserved. For example, the orientation of these regions and their spacers at the Ig λ locus is inverted relative to that at the Ig κ locus (where the regions with the 12-nucleotide spacer are adjacent to the V regions). These relationships are summarized in the table below.

	Length of Spacer in Joining Sequence (nt)		
Locus	V Segment	D Segment	J Segment
Ig heavy chain	23	12 — 12	23
Ig κ	12		23
Ig λ	23		12
T cell α	23		12
T cell δ	23	12 — 23	12
T cell β	23	12 — 23	12
T cell γ	23		12

The sequences listed above are not absolutely conserved. The following figure shows the distribution of the nucleotides at each position around the region of the joining elements adjacent to the Ig λ V regions, which contain the 23-nucleotide spacer. Darker boxes indicate a higher level of sequence conservation. The 7-mer is at positions 7 to 13, and the 9-mer is at positions 37 to 45. Note the additional highly conserved positions in the 23-nucleotide spacer region.

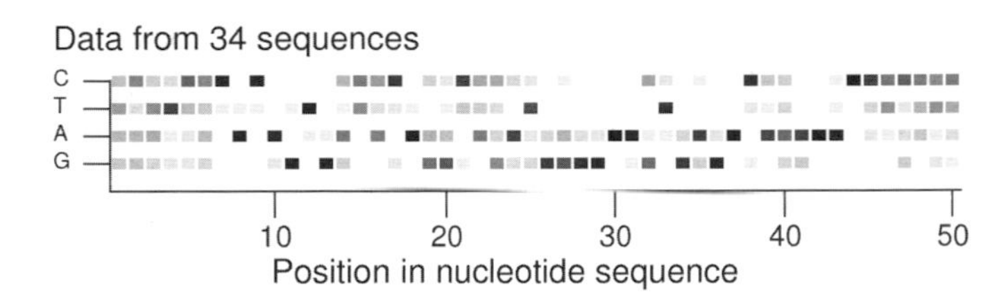

The following figure shows similar data from the sequences adjacent to the T-cell receptor β locus V genes. These sequences also have the 23-nucleotide spacing of the conserved elements. Note the more limited conservation in this set and the rather different sequences in the spacer.

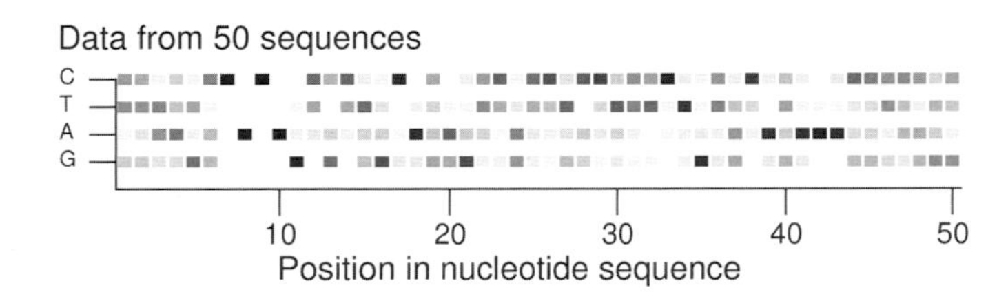

Sequences with the 23-nucleotide spacer (shown above) are joined to elements with the 12-nucleotide spacer. The figure below shows the sequences from the 12-nucleotide spacer regions that are used to direct joining at the amino-terminal end of the immunoglobulin heavy chain D segments. The 9-mer is at positions 17 to 25, and the 7-mer is at positions 38 to 44.

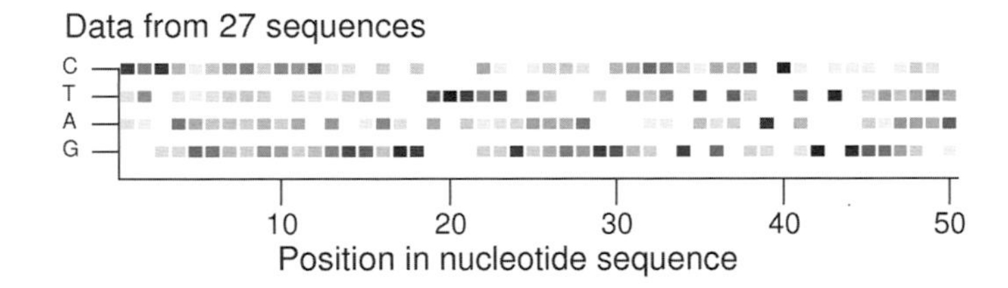

Data Sources, Methods, and References

The nucleotide sequences used to generate the first figure were from 34 of the 38 Ig λ V regions annotated as genes (not pseudogenes) onto release 36.2 of the reference genome sequence. The remaining four sequences did not retain the precise 23-nucleotide spacing between the consensus sequences shown on the previous page or did not preserve the precise position of the 7-mer relative to the annotated end of the V gene segment. The second figure used 50 of 51 annotated segments (the one inverted V was excluded); the third figure used all 27 of the D segments.

See also:

Akamatsu Y. et al. 1994. Essential residues in V(D)J recombination signals. *J. Immunol.* **153:** 4520–4529.

Akira S. et al. 1987. Two pairs of recombination signals are sufficient to cause immunoglobulin V-(D)-J joining. *Science* **238:** 1134–1138.

Olaru A. et al. 2005. Beyond the 12/23 rule of VDJ recombination independent of the Rag proteins. *J. Immunol.* **174:** 6220–6226.

CHAPTER NINE

POLYMORPHISM

MUCH OF OUR KNOWLEDGE OF THE HUMAN GENOME IS BUILT from a collection of reference sequences. In this chapter, the focus shifts to variation observed among individuals. Some of the sections in this chapter describe the specific alleles present in the reference genome, highlighting the kinds of information one can easily obtain from the larger-scale sequencing of individuals that is likely in the near future.

Although a significant part of the observed polymorphism is essentially neutral from the perspective of selection, many loci have been studied because of their association with specific genetic diseases. Once a disease-associated allele becomes more frequent in a population, the question of positive selection may arise, often in the context of resistance to some infectious agent. Host defense is a factor in the extraordinary level of polymorphism at the major histocompatibility locus.

Also described in this chapter is an important class of variation within an individual: the tumor suppressor loci that are inactivated in various cancers. Many other loci encode oncogenes that are related to genes in retroviruses. Those are discussed in chapter 10.

What Is the Frequency of Single Nucleotide Polymorphisms in the Genome?

Single nucleotide polymorphisms (SNPs) play a central role in population genetics studies and in the search for mutations associated with specific genetic diseases. Estimates of their frequency provide a baseline for identification of particular alleles that may have elevated or reduced frequencies associated with selective pressures. The values in the table below are from the large-scale study by The International SNP Map Working Group. They provide an estimate of the frequency of heterozygous positions in pairs of chromosomes in the population.

Chromosomes	SNPs/kb in Chromosome Pairs
All	0.751
X	0.469
Y	0.151

As can be seen from the table, the frequency of SNPs on the sex chromosomes is reduced. This is consistent with population genetics models. Note that the value for the Y chromosome excludes the pseudoautosomal regions.

Data Sources, Methods, and References

Sachidanandam R. et al. 2001. A map of human genome sequence variation containing 1.42 million single nucleotide polymorphisms. *Nature* **409:** 928–933.

Which ABO Allele Is in the Reference Genome?

One of the most widely studied polymorphic loci is the ABO locus, which controls the familiar blood types. The reference human genome sequence has an O allele at this locus. Shown below are the results of a search of the reference genome sequence with the RefSeq messenger RNA (mRNA) sequence for ABO (an A allele) as query. The amino acid translations of the sequences in the relevant reading frames are also shown.

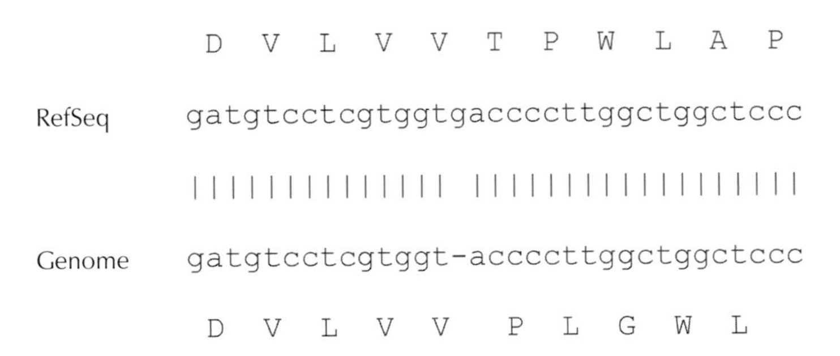

If the function of the ABO locus was not known, its annotation in the reference genome could have been quite different. The O allele contains a frameshift at codon 87 that translates a 117-aa protein (see additional details below). In contrast, A and B alleles each encode 354-aa proteins. The O allele has three differences relative to an A sequence upstream of codon 87: V36F, R63H, and P74S. Additional alignments of the sequences above show these three expected substitutions (not shown).

Some alleles at the ABO locus encode functional sugar transferases. The product of an A allele adds an *N*-acetylgalactosamine moiety, and the product of a B allele adds a galactose moiety. There are four amino acid differences between the products of the RefSeq (A) and B alleles that are referenced below.

Data Sources, Methods, and References

The RefSeq mRNA for the ABO locus is GI:58331215 (this is the sequence of an A allele, but the header for this RefSeq entry describes both enzymatic activities for this locus). The CDS region of this sequence was used to search the ABO region of the reference genome sequence with NCBI BLASTN. The 117-aa protein encoded by an O allele can be found in GenBank GI:226864. An example of the protein sequence encoded by a B allele is GI:226863.

Similar searches of the alternate assembly based on the Celera sequence do not have the frameshift.

See also:

Yamamoto F. et al. 1990. Molecular genetic basis of the histo-blood group ABO system. *Nature* **345:** 229–233.

How Variable Are the HLA Proteins?

Among the most variable of human proteins are those of the major histocompatibility complex. Many full-length protein sequences are available for the highly polymorphic class I genes HLA-A, HLA-B, and HLA-C. The HLA-A, HLA-B, and HLA-C alleles in the reference genome sequence are the same as those in RefSeq (there is only one allele for each locus in RefSeq). They are HLA-A*3, HLA-B*7, and HLA-Cw*7, respectively.

The table below gives some basic statistical information about the levels of variation at the HLA-A, HLA-B, and HLA-C loci.

Locus	Size (aa)	No. of Sequences	Average No. of Amino Acid Differences (Pairwise)	No. of Variable Amino Acids
HLA-A	365	30	23.9	79
HLA-B	362	54	21.2	62
HLA-C	366	18	21.4	72

Based on the average SNP levels in the genome, one would expect very few differences in pairwise comparisons of alleles for proteins of this size. A large number of sequences were used to produce the table, and there are a large number of differences between many of them.

As shown on the next page, the variable positions are not randomly distributed along the length of the sequences. Some positions have notable levels of polymorphism while other positions are more conserved. At most variable positions, only a second amino acid is found; however, at some positions in the protein, five or six different amino acids are observed. Note the higher levels of conservation closer to the carboxy-terminus of the protein.

The following figure presents the number of amino acids observed at each position in the set of 54 HLA–B proteins.

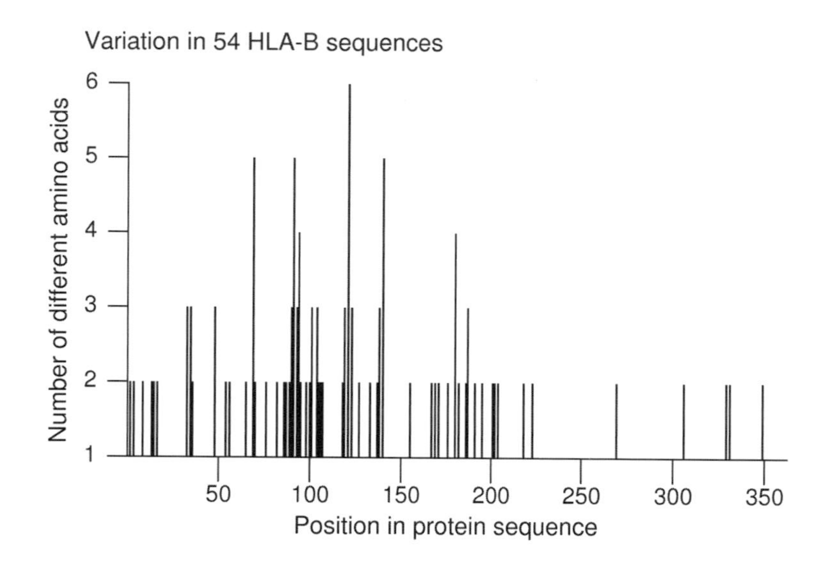

The figure below shows which amino acids are found in the more polymorphic region of HLA-B. The segment presented is residues 69 to 140, and it corresponds to the portion of the figure above where five or more amino acids are seen at some positions. Note the nonconservative substitutions at the most variable residues. In the figure, darker boxes indicate a higher fraction of use of that amino acid at a given position.

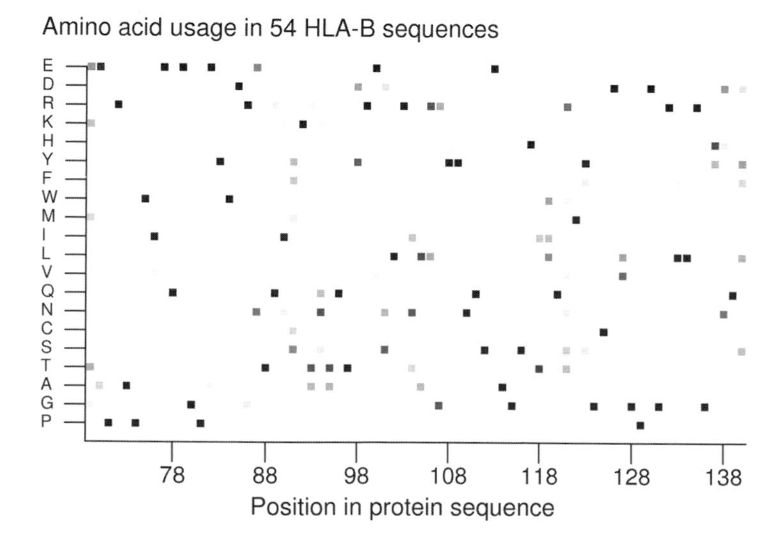

Data Sources, Methods, and References

Different HLA protein sequences were selected from the GenBank entries for HLA-A, HLA-B, or HLA-C. All of the sequences used were the length indicated with no ambiguous residues. The set of HLA-B sequences used in both figures was the same as that in the table.

The identities of the RefSeq entries were determined with BLASTP. No sequence differences were found between the RefSeq entries and the HLA-A*3, HLA-B*7, and HLA-Cw*7 alleles. The HLA alleles are annotated on the reference genome sequence as A*030101, B*070201, and Cw*07020101, respectively.

Which Genes Have Common Polymorphisms?

Aside from the ABO blood group and HLA proteins, which are described separately on pages 135–139, a number of human genes have well-characterized, high-frequency polymorphisms. Several of these genes are described in the table below.

Gene	Protein/Function	Common Polymorphisms and Their Distribution
CCR5[1]	chemokine receptor	32-bp deletion associated with resistance to HIV-1; ~10% in Europeans, lower in others
TAS2R38[2]	PTC (bitter) taste receptor	two common alleles outside of Africa differ at three amino acids; additional alleles in Africa
DARC[3]	Duffy blood group	alleles with geographical restriction; promoter mutation provides resistance to *P. vivax* malaria
LCT[4]	lactase	promoter mutation leading to lactase persistence at very high frequency in Northern Europe; other promoter polymorphisms in Africa

The footnotes are in "Data Sources, Methods, and References."

Data Sources, Methods, and References

The following references are cited in the table:

[1]Martinson J.J. et al. 1997. Global distribution of the *CCR5* gene 32-basepair deletion. *Nat. Genet.* **16:** 100–103.

[2]Kim U.K. et al. 2003. Positional cloning of the human quantitative trait locus underlying taste sensitivity to phenylthiocarbamide. *Science* **299:** 1221–1225.

[3]Carter R. and Mendis K.N. 2002. Evolutionary and historical aspects of the burden of malaria. *Clin. Microbiol. Rev.* **15:** 564–594.

[4]Ingram C.J. et al. 2007. A novel polymorphism associated with lactose tolerance in Africa: Multiple causes for lactase persistence? *Hum. Genet.* **120:** 779–788.

Which Genes and Alleles Are Associated with Common Genetic Diseases?

Some well-known genetic diseases are caused by a common allele. In other cases, there is a great diversity of alleles. Certain diseases have become associated with a particular allele (such as sickle cell anemia) but other mutant alleles of that locus may be present in the population at comparable frequencies. The table below gives some examples of each type. In cases of incomplete penetrance, the most common allele may not be the most commonly associated with disease.

Disease	Gene	Alleles[8]
Breast/Ovarian Cancer	*BRCA1*[1]	frameshifts 185ΔAG and less-frequent 5382insC common in Ashkenazi Jews; many others known
Breast/Ovarian Cancer	*BRCA2*[1]	frameshift 6174ΔT common in Ashkenazi Jews; many others known
Cystic Fibrosis	*CFTR*[2]	~70% of mutations are ΔF508 but many others are known
Duchenne/Becker Muscular Dystrophy	*DMD*[3]	often deletions, mutations that preserve reading frame more common in Becker type
G6PDH Deficiency	*G6PD*[4]	many alleles known, S188F (Med) is common in Southern Europe, V68M + N126D (A-) is common in Africa
Sickle Cell Anemia	*HBB*[5]	HbS is E6V, HbC is E6C; these alleles are found in comparable frequencies in parts of Burkina Faso
Hemochromatosis	*HFE*[6]	C282Y common among Northern European patients; others include H63D
Mediterranean Fever	*MEFV*[7]	several alleles including M694V

The footnotes are in "Data Sources, Methods, and References."

Data Sources, Methods, and References

Note that several of the genes in the table produce multiple transcripts and isoforms, and the alteration of interest may not be present at the indicated location in all transcripts or isoforms.

The following are the footnotes to the table above:

[1]Phelan C.M. et al. 2002. A low frequency of non-founder *BRCA1* mutations in Ashkenazi Jewish breast-ovarian cancer families. *Hum. Mutat.* **20:** 352–357.

[2]Zielenski J. and Tsui L.C. 1995. Cystic fibrosis: genotypic and phenotypic variations.

Annu. Rev. Genet. **29:** 777–807.

[3]Aartsma-Rus A. et al. 2006. Entries in the Leiden Duchenne muscular dystrophy mutation database: An overview of mutation types and paradoxical cases that confirm the reading-frame rule. *Muscle Nerve* **34:** 135–144.

[4]Beutler E. 1996. *G6PD*: Population genetics and clinical manifestations. *Blood Rev.* **10:** 45–52.

[5]Modiano D. et al. 2001. Haemoglobin C protects against clinical *Plasmodium falciparum* malaria. *Nature* **414:** 305–308.

[6]Merryweather-Clarke A.T. et al. 1997. Global prevalence of putative haemochromatosis mutations. *J. Med. Genet.* **34:** 275–278.

[7]The International FMF Consortium. 1997. Ancient missense mutations in a new member of the RoRet gene family are likely to cause familial Mediterranean fever. *Cell* **90:** 797–807.

[8]Nucleotide changes begin with a number; otherwise, protein changes are listed.

Which Genes Function as Tumor Suppressors?

A large number of genes have been identified as tumor suppressors, and these genes may contain polymorphisms that lead to disease. As can be seen with the genes listed in the table below, tumor suppressors function in diverse regulatory pathways in cells. Many of the genes in the table have numerous synonyms, but only the current standard names are listed.

Gene	Protein/Function
APC[1]	adenomatosis polyposis coli, Wnt signals
CADM1[2]	cell adhesion protein
CDH1[3]	epithelial cadherin
CDKN2A[4]	INK4a/b and ARF proteins, cell cycle control
CHFR[5]	ring finger protein, mitotic checkpoint
DAPK1[6]	protein kinase, cell death pathways
DCC[7]	netrin receptor, cell death
EPB41L3[2]	cytoskeletal protein
NF1[8]	neurofibromin, Ras signals
NF2[8]	merlin, cytoskeletal linker protein
PTEN[9]	IP3/protein phosphatase
RB1[10]	retinoblastoma protein, cell cycle control
STK11[11]	protein kinase
TP53[12]	p53, transcriptional regulation
TSC1[13]	hamartin, complexes with TSC2
TSC2[13]	tuberin, complexes with TSC1, GTPase activator
VHL[14]	Von Hippel-Lindau protein, ubiquitin ligase
WT1[15]	Wilms tumor protein

The footnotes are in "Data Sources, Methods, and References."

Not included in the table are a number of genes with functions in DNA repair processes. Loss-of-function mutations in such genes as *BRCA1*, *BRCA2*, *MLH1*, *MSH2*, *MSH6*, and *MUTYH* are also associated with the development of tumors in different tissues.

Data Sources, Methods, and References

For additional information on DNA repair functions, see:

David S.S. et al. 2007. Base-excision repair of oxidative DNA damage. *Nature* **447:** 941–950.

Narod S.A. and Foulkes W.D. 2004. *BRCA1* and *BRCA2*: 1994 and beyond. *Nat. Rev. Cancer* **4:** 665–676.

Rowley P.T. 2005. Inherited susceptibility to colorectal cancer. *Annu. Rev. Med.* **56:** 539–554.

The following references are cited in the table:

[1]Bienz M. 2002. The subcellular destinations of APC proteins. *Nat. Rev. Mol. Cell Biol.* **3:** 328–338.

[2]Yageta M. et al. 2002. Direct association of TSLC1 and DAL-1, two distinct tumor suppressor proteins in lung cancer. *Cancer Res.* **62:** 5129–5133.

[3]Guilford P. et al. 1998. E-cadherin germline mutations in familial gastric cancer. *Nature* **392:** 402–405.

[4]Gil J. and Peters G. 2006. Regulation of the INK4b-ARF-INK4a tumour suppressor locus: All for one or one for all. *Nat. Rev. Mol. Cell Biol.* **7:** 667–677.

[5]Scolnick D.M. and Hazalonetis T.D. 2000. *Chfr* defines a mitotic stress checkpoint that delays entry into metaphase. *Nature* **406:** 430–435.

[6]Bialik S. and Kimchi A. 2006. The death-associated protein kinases: structure, function, and beyond. *Annu. Rev. Biochem.* **75:** 189–210.

[7]Furne C. et al. 2006. The dependence receptor DCC requires lipid raft localization for cell death signaling. *Proc. Natl. Acad. Sci.* **103:** 4128–4133.

[8]Gutmann D.H. 2001. The neurofibromatoses: when less is more. *Hum. Mol. Genet.* **10:** 747–755.

[9]Goberdham D.C. and Wilson C. 2003. PTEN: tumour suppressor, multifunctional growth regulator and more. *Hum. Mol. Genet.* **12:** R239–R248.

[10]Claudio P.P. et al. 2002. The retinoblastoma family: twins or distant cousins? *Genome Biol.* **3:** 3012.

[11]Jenne D.E. et al. 1998. Peutz-Jeghers syndrome is caused by mutations in a novel serine threonine kinase. *Nat Genet.* **18:** 38–43.

[12]Vousden K.H. and Lane D.P. 2007. p53 in health and disease. *Nat. Rev. Mol. Cell Biol.* **8:** 275–283.

[13]Kwiatkowski D.J. and Manning B.D. 2005. Tuberous sclerosis: a GAP at the crossroads of multiple signaling pathways. *Hum. Mol. Genet.* **14:** R251–R258.

[14]Ohh M. 2006. Ubiquitin pathway in VHL cancer syndrome. *Neoplasia.* **8:** 623–629.

[15]Wagner K.D. et al. 2003. The complex life of WT1. *J. Cell Sci.* **116:** 1653–1658.

CHAPTER TEN

COMPARATIVE GENOMICS

THE HUMAN GENOME IS STUDIED IN THE CONTEXT OF OUR KNOWLEDGE about other organisms. Homologs of human genes are generally readily identified in many other species, sometimes in species not expected to share those functions. In some cases, these efforts focus on the similarities and differences of these homologs in other mammals or in other vertebrates. Other work concentrates on linkages to more distantly related but intensively studied organisms such as *D. melanogaster*, *C. elegans*, and *S. cerevisiae*. Some studies focus on relationships that extend beyond the eukaryotes. Another important area of investigation is the similarity of human genes to those found in viruses that infect humans or other vertebrates.

This chapter attempts to provide background information on the use of comparative genomics with human DNA and protein sequences. Rather than presenting results from comprehensive sequence comparisons, examples have been chosen to represent the broad range of sequence conservation across species. Some of these examples highlight complications in such analyses. Areas considered include genome size, the rate of evolution of individual genes, and changes in the size and structure of gene families. Other topics include the presence or absence of proteins with specific functions in different species and how human genes are related to genes in eubacteria and archaea.

What Are the Major Differences Between the Human and Mouse Genomes?

Perhaps the most important species used for comparative genomics studies with humans is the mouse. Although it is relatively straightforward to identify the mouse counterparts of most human genes and the degree of sequence similarity between the two species is generally quite high, there are some significant differences in the overall genomes of the two species. Some of these differences are highlighted in the table below.

Feature	Human	Mouse
No. of chromosomes	22 + XY	19 + XY
Sequenced DNA	2858 Mb	2559 Mb
Largest autosome (total size)	247 Mb	197 Mb
Smallest autosome (total size)	47 Mb	61 Mb
X chromosome (total size)	155 Mb	167 Mb

The total amount of sequenced DNA in the mouse genome is more than 10% smaller than the total amount of sequenced DNA in the human genome (see the end of this section for a discussion of the unsequenced DNA). In both species, the chromosomes named "1" are the largest. The smallest autosomes are human chromosome 21 and mouse chromosome 19. The mouse X chromosome is estimated to be slightly larger than its human counterpart. Comparing the sizes of the Y chromosomes is complicated by the large fraction of repeats.

Data Sources, Methods, and References

The numbers in the table are from the assembled sequences of the NCBI reference genomes. For humans, release 36.2 of the annotated genome was used. For mice, it was release 37. Single-base ambiguities and large blocks of Ns, which estimate unsequenced segments, were excluded from the totals in the "Sequenced DNA" row. The unsequenced DNA in the assembled chromosomes totals 222 Mb in humans and 96 Mb in mice.

For the chromosomes listed in the last three rows of the table, the "total size" includes the sequenced and unsequenced fractions. Because of uncertainty from unsequenced regions, the chromosome sizes for both species are estimates. This is especially true for the Y chromosomes. For mice, the Y reference sequence is only 15.9 Mb, 13.2 Mb of which is composed of blocks of Ns. The NCBI Map Viewer draws the mouse Y chromosome similar in size to mouse chromosome 19.

With the Y chromosomes excluded, the total genome sizes, including unsequenced regions, would be 3023 Mb and 2639 Mb for humans and mice, respectively, and the mouse genome would be about 87% the size of the human genome. With the Y chromosomes included, the difference is likely to be somewhat smaller.

Both species also have small amounts of sequences that are not located on the chromosomes. These sequences are not included in the totals above. The mitochondrial genomes were also excluded.

How Do the Mitochondrial Genomes Vary Among Species?

Mitochondrial genomes vary considerably among the eukaryotes in both size and gene complement. This information is summarized in the table below.

Species	Size (bp)	Comments
Human	16,571	alternate reference sequence is 16,569 bp
Mouse	16,299	
Chicken	16,775	
Zebrafish	16,596	
Sea urchin	15,650	different genetic code
D. melanogaster	19,517	different genetic code and gene order
C. elegans	13,794	different genetic code and gene order
A. thaliana	366,924	many more genes, including some for ribosomal proteins; uses the standard genetic code
S. cerevisiae	85,779	many more genes, but lacks the NADH dehydrogenase subunits; different genetic code

Although vertebrate mitochondrial genomes are generally regarded as compact, the sea urchin genome is about 1 kb smaller and encodes the same genes. The *C. elegans* mitochondrial genome is even smaller (see additional details below).

The mitochondrial genome of *D. melanogaster* is larger than those of the other animals shown in the table, but other *Drosophila* species are close to the size of vertebrate mitochondria (not shown).

Plant and fungal mitochondrial genomes are generally larger than animal mitochondrial genomes. Although *S. cerevisiae* lacks the genes for the NADH dehydrogenase subunits, these genes are present in many other fungi such as *C. albicans* or *A. niger* (not shown).

Data Sources, Methods, and References

In addition to the human mitochondrial reference sequences (GI:17981852 and the alternate sequence GI:115315570), the mitochondrial sequences used were GI:34538597 (mouse, C57BL/6J), GI:5834843 (chicken), GI:15079186 (zebrafish), GI:5834897 (sea urchin), GI:5835233 (*D. melanogaster*), GI:5834884 (*C. elegans*), GI:26556996 (*A. thaliana*), and GI:6226515 (*S. cerevisiae*).

In the table above, the "standard genetic code" refers to NCBI translation table 1 (mitochondria from *A. thaliana* use the same code as human nuclear genes), and the "different genetic code" refers to a code other than the vertebrate mitochondrial code (NCBI translation table 2).

The annotated reference sequence for the *C. elegans* mitochondrial genome does not include the small *ATP8* gene. If *C. elegans* encodes an *ATP8* gene, its sequence is quite different from that of the human *ATP8* gene.

How Does Codon Usage Differ in the Eukaryotes?

Codon usage patterns differ significantly among the eukaryotes. This is illustrated in the table below, where the codon usage patterns for human (*H. sapiens* [Hs]) ACTB (cytoplasmic actin) are compared with the closely related ACT5C and ACT1 proteins from *D. melanogaster* (Dm) and *S. cerevisiae* (Sc), respectively.

Codon	Amino Acid	Hs	Dm	Sc	Codon	Amino Acid	Hs	Dm	Sc	Codon	Amino Acid	Hs	Dm	Sc	Codon	Amino Acid	Hs	Dm	Sc
UUU	F	1	2	2	UCU	S	2	3	14	UAU	Y	3	4	0	UGU	C	2	2	4
UUC	F	12	10	12	UCC	S	16	12	12	UAC	Y	12	12	14	UGC	C	4	5	0
UUA	L	0	0	2	UCA	S	0	0	3	UAA	*	0	1	1	UGA	*	0	0	0
UUG	L	0	3	19	UCG	S	1	7	0	UAG	*	1	0	0	UGG	W	4	4	4
CUU	L	0	1	2	CCU	P	6	1	3	CAU	H	1	1	3	CGU	R	4	12	5
CUC	L	6	3	0	CCC	P	12	13	7	CAC	H	8	8	7	CGC	R	7	6	0
CUA	L	1	0	2	CCA	P	1	2	14	CAA	Q	1	0	14	CGA	R	0	0	0
CUG	L	20	21	0	CCG	P	0	3	0	CAG	Q	11	12	0	CGG	R	4	0	0
AUU	I	4	4	14	ACU	T	5	0	0	AAU	N	2	1	0	AGU	S	0	0	2
AUC	I	24	23	16	ACC	T	17	24	9	AAC	N	7	8	9	AGC	S	6	3	0
AUA	I	0	0	0	ACA	T	2	1	6	AAA	K	3	0	6	AGA	R	1	0	13
AUG	M	17	16	16	ACG	T	2	0	12	AAG	K	16	19	12	AGG	R	2	0	0
GUU	V	1	4	14	GCU	A	7	9	11	GAU	D	5	9	11	GGU	G	2	12	29
GUC	V	8	6	12	GCC	A	18	18	9	GAC	D	18	12	9	GGC	G	23	11	0
GUA	V	0	0	0	GCA	A	2	2	26	GAA	E	2	3	26	GGA	G	0	5	0
GUG	V	13	13	0	GCG	A	2	1	1	GAG	E	24	25	1	GGG	G	3	0	0

*Indicates a stop codon. For the 1-letter amino acid code, see page 70.

Codon usage in *D. melanogaster* is generally more similar to codon usage in humans than in yeast. Some of the clearest differences between humans and yeast are seen with the glutamine, glutamate, and glycine codons.

Data Sources, Methods, and References

The sequences used to generate the codon usage table in this section were the CDS regions in GI:5016088 (ACTB), GI:24639996 (ACT5C), and GI:42742172 (ACT1).

How Similar Are Human Proteins to Those in Other Species?

Different types of human proteins have quite different degrees of sequence similarity to their counterparts in other species. Some examples are shown in the figure below.

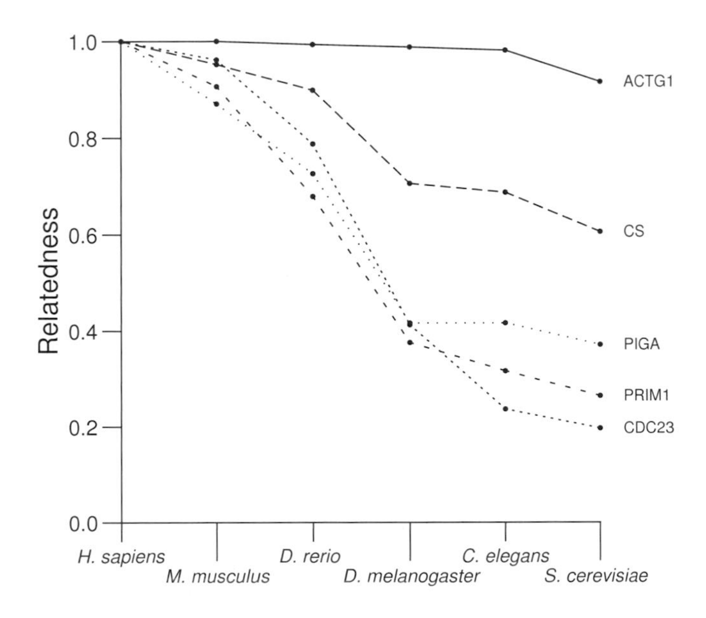

The proteins used were ACTG1 (γ1 actin), CS (citrate synthase), PIGA (in the GPI anchoring pathway), PRIM1 (DNA primase subunit), and CDC23 (anaphase promoting complex subunit). These proteins produce relatively straightforward plots related to evolutionary distance and the intrinsic conservation of protein function. Some examples presented in later sections are more complex (see pp. 157 and 160).

Just as related sequences in other species may perform different functions, the absence of related sequences in other species does not indicate that the other species lacks those functions. This is readily shown with the enzymes of the glycolytic pathway, which are some of the most widely distributed metabolic enzymes. Although the counterparts for human glycolytic enzymes can be easily identified in other vertebrates and in *Drosophila*, the corresponding enzymes in other species frequently have unrelated sequences.

The table on the following page summarizes the results of BLASTP searches of proteins from *C. elegans* and selected microbial species starting with human sequences for the glycolytic enzymes. In the majority of cases, a related sequence is readily identified ("yes" in the table). In each species, at least one of the glycolytic enzymes is not readily found. However, these species do have enzymes for the steps (see the details at the end of this section). The match to glyceraldehyde-3-phosphate dehydrogenase (GAPDH) in *A. pernix* is relatively weak and might not be identified without additional sequence comparisons beyond that with the human GAPDH sequence.

Human Enzyme	*C. elegans*	*S. cerevisiae*	*E. coli*	*A. pernix*
glucose phosphate isomerase	yes	yes	yes	no
phosphofructokinase	yes	yes	yes	no
aldolase	yes	no	no	no
triosephosphate isomerase	yes	yes	yes	no
glyceraldehyde-3-phosphate dehydrogenase	yes	yes	yes	very weak
phosphoglycerate kinase	yes	yes	yes	yes
phosphoglycerate mutase	no	yes	yes	no
enolase	yes	yes	yes	yes
pyruvate kinase	yes	yes	yes	yes

Several human glycolytic enzymes are encoded by small gene families, but their members are closely related and do not affect the results significantly. The sperm-expressed GAPDHS protein produces even weaker matches than GAPDH in *A. pernix*. Small gene families are sometimes found in the comparative species.

There are many differences in the glycolytic enzymes of eukaryotes and archaea. In later sections, a number of important similarities between eukaryotes and archaea are described (see pp. 154–156).

Data Sources, Methods, and References

The method used to generate the points for the plot began with HSP scores from BLASTP (from NCBI BLAST 2.211). The *y*-axis is the ratio of the BLASTP score obtained with the best-matching protein in another species to the score from the self-match (both scores were adjusted as described in chapter 1). The scale on the *x*-axis is arbitrary and is not related to any measure of evolutionary relatedness.

The sequences for the enzymes in the table that were not readily identified via BLASTP are as follows: GI:118431188 (*A. pernix* phosphoglucose/phosphomannose isomerase), GI:118431733 (likely *A. pernix* phosphofructokinase, note also GI:14600388), GI:6322790 (*S. cerevisiae* aldolase), GI:90111385 (*E. coli* aldolase class I; this species also has a class II enzyme), GI:118430840 (*A. pernix* aldolase class I; this species may also have a class II aldolase), GI:118431499 (*A. pernix* triose phosphate isomerase), GI:17507741 (*C. elegans* phosphoglycerate mutase), and GI:118431534 (*A. pernix* phosphoglycerate mutase).

How Similar Are Human Genes to Those in Other Species at the Nucleotide Level?

Gene comparisons can be made at the nucleotide level as well as at the protein level. These comparisons can provide additional information such as the relative rates of synonymous and nonsynonymous substitutions. The figure below presents some sample comparisons using the same genes as used for the protein-level examples on page 151. Unlike the values in the protein comparisons, these are simple percent identity values.

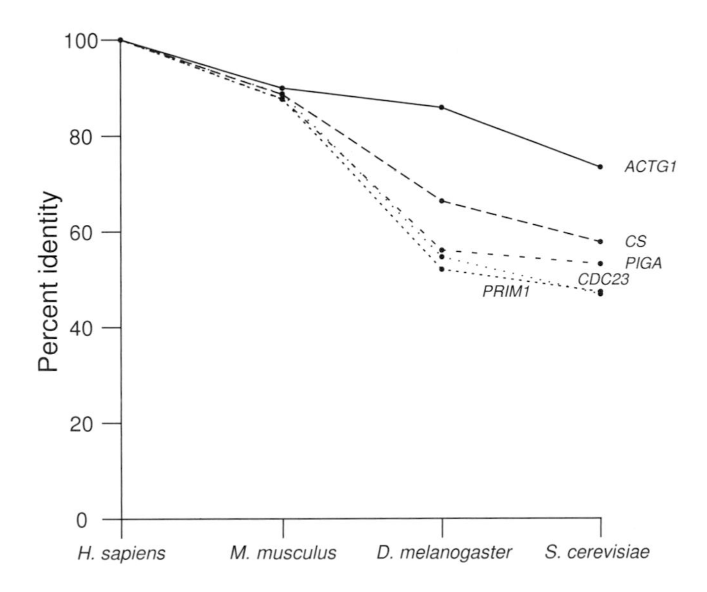

Nucleotide sequences can evolve more rapidly than protein sequences. For mice, there is less relative variation among these genes at the nucleotide level than at the protein level. Note that more positions are being compared at the nucleotide level than at the protein level. Nucleotide sequences also have a higher level of chance matches because of the limited number of possibilities at each position. So for distant species such as yeast, the percent identity remains relatively high.

Data Sources, Methods, and References

The values used to generate the figure are from the NCBI HomoloGene database. They are not adjusted for multiple substitutions. HomoloGene is part of the NCBI Entrez system. Its home page can be found at http://www.ncbi.nlm.nih.gov/sites/entrez?db=homologene.

How Are Human and Bacterial Ribosomes Related?

The human genome encodes two complete translation systems, one in the cytoplasm and a second in the mitochondria. Only a few ribosomal proteins in the two systems have readily detectable sequence similarity. These pairs include: RPL8 and MRPL2, RPS14 and MRPS11, and RPS23 and MRPS12 (names beginning with the letter M are nuclear-encoded mitochondrial ribosomal proteins). These pairs have similarity to proteins in both eubacteria and archaea. The figure below shows these relationships.

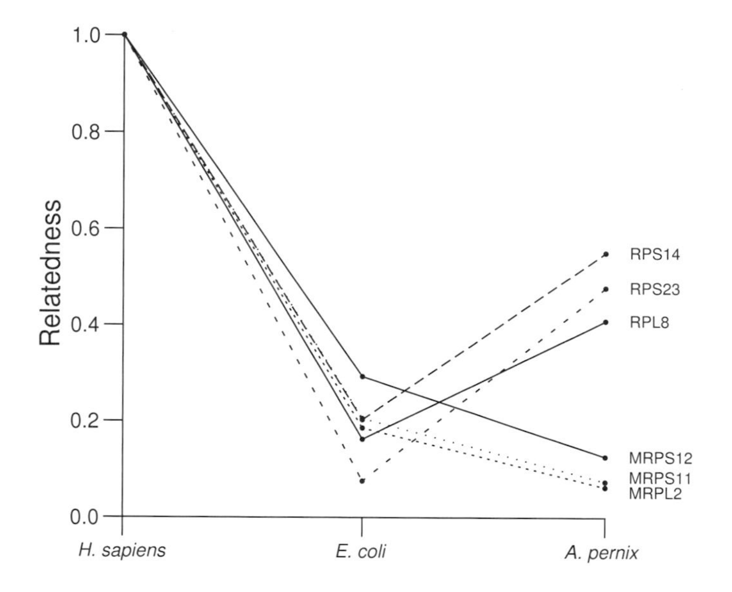

In all cases, the mitochondrial proteins have greater similarity to proteins in *E. coli* while the cytoplasmic ribosomal proteins are much more closely related to the proteins of the crenarchaeon *A. pernix*. Similar results can be obtained with cytoplasmic and mitochondrial aminoacyl-transfer RNA (tRNA) synthetases and translation factors (not shown).

Data Sources, Methods, and References

The relationships between the cytoplasmic and mitochondrial human ribosomal proteins were determined using NCBI BLAST 2.2.11. The searches of the protein sets were also performed with NCBI BLAST 2.2.11. The query sequences were GI:5032051 (RPS14), GI:4506701 (RPS23), GI:4506663 (RPL8), GI:16950591 (MRPS12), GI:16554609 (MRPS11), and GI:21265070 (MRPL2). The graph was generated using the approach described in chapter 1.

Are Human DNA Replication Proteins Related to Those of Bacteria?

Although there are many differences between the DNA replication systems of bacteria and eukaryotes, many similarities can be seen between the human replication proteins and those of both eubacteria and archaea. Some of these are shown in the following two figures. It is important to note that the two species used here are not completely representative of eubacteria and archaea and that some exceptions to the patterns presented here can be found.

As described on page 100, human DNA polymerases can be grouped on the basis of sequence similarity. The figure below shows results with some representative human DNA polymerases.

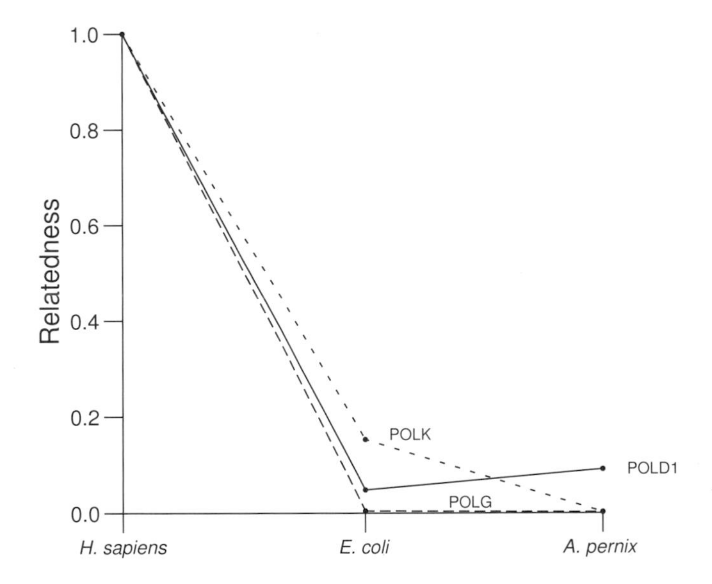

DNA polymerases are generally large proteins with only limited similarity to bacterial enzymes. Interestingly, POLG, the mitochondrial DNA polymerase, is not related to proteins in *E. coli* or *A. pernix*. Similar results are seen with the POLB family.

Members of the POLD1 group (including POLA) are more closely related to archaeal types. In contrast, the POLK family (and the POLN family, not shown) has more similarity to eubacterial enzymes.

Human DNA replication requires many other proteins besides the DNA polymerases. Many of these proteins form large complexes. Some of these proteins are presented in the figure on the next page.

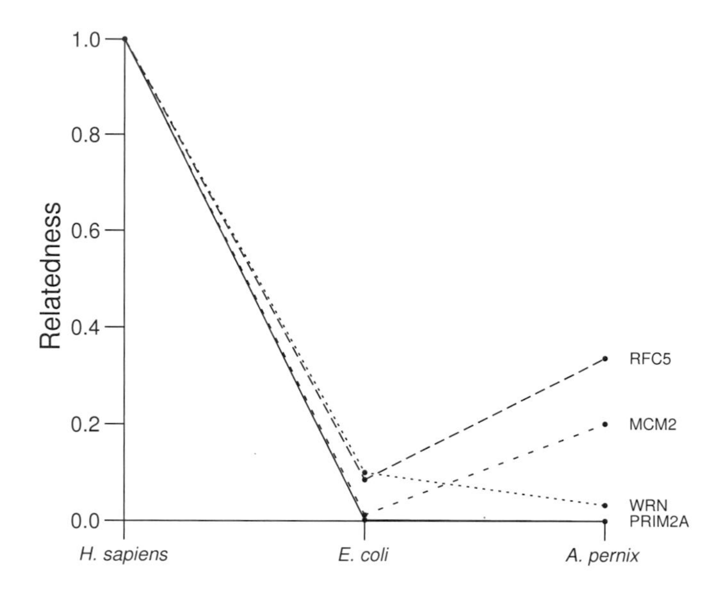

RFC5 (a component of replication factor C) and MCM2 (in the MCM complex) both show greater similarity to archaeal proteins than to eubacterial proteins. Similar results are obtained with ORC1L (origin recognition complex subunit; not shown). The human genome encodes three DNA ligases that give similar results (also not shown). As seen in the figure above, WRN (Werner helicase) gives the opposite result.

Homologs of PRIM2A (the large subunit of DNA primase) are not readily detected in *E. coli* or in *A. pernix*. Similar results (not shown) are found with the small subunit of primase and with the proteins of the GINS complex.

Data Sources, Methods, and References

Both figures were generated by the method described in chapter 1. All of the searches used NCBI BLAST 2.2.11. The query sequences were GI:41349495 (PRIM2A), GI:6677723 (RFC5), GI:33356547 (MCM2), and GI:110735439 (WRN).

It is important to note that some of the results with specific enzymes can derive from mechanistic differences. For example, *E. coli* DNA ligase uses NAD rather than ATP as its cofactor. See:

Wilkinson A. et al. 2001. Bacterial DNA ligases. *Mol. Microbiol.* **40:** 1241–1248.

Are Human Proteins for mRNA Processing Functions Conserved in Eukaryotes?

The human genome encodes many proteins involved in the processing of messenger RNAs (mRNAs). Some of these proteins are readily identified in diverse eukaryotes; however, others are not. Sequence similarity is one way to identify related proteins in other species. In the figure below, sequence similarity was used to examine mRNA processing in some widely used eukaryotic model systems.

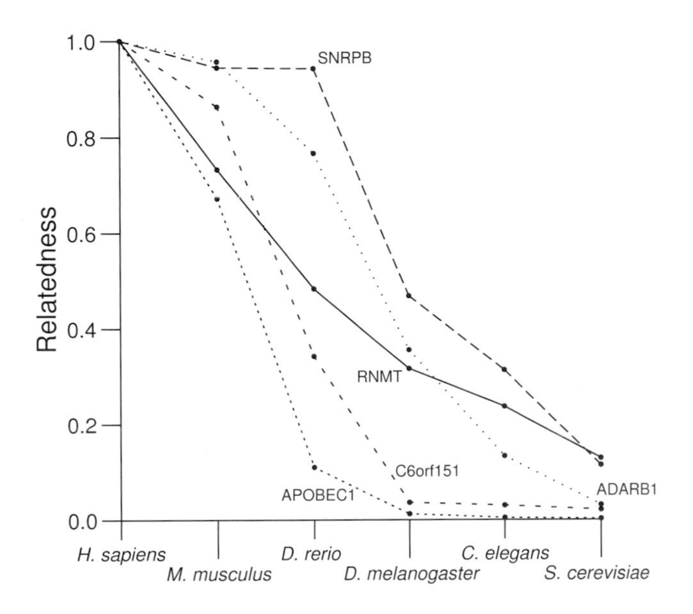

RNMT is the capping methyltransferase. SNRPB is one of the snRNP core proteins. Proteins that perform these functions are present in all of the species shown; however, their sequence conservation pattern is somewhat different. Humans have a second protein that is closely related to SNRPB.

C6orf151 is a protein in the minor spliceosome. *Drosophila* does have a diverged U11/U12-type splicing system.

APOBEC1 is in the cytosine-to-uracil RNA editing complex. It is part of a family of cytosine deaminases with diverse functions. The match in zebrafish to APOBEC1 is actually the counterpart of a different protein, APOBEC2. ADARB1 is involved in adenine-to-inosine RNA editing and is also in a gene family. The weak yeast match is to a tRNA adenosine deaminase.

Sequence similarity is not always a straightforward way to test for the presence of functionally related pathways in different species. Many of these genes are in families whose members have very different functions. Sometimes the matching protein in the other species can be used in a reverse search to identify a better match in humans, thus providing clues to its function. A second issue is that a protein that is unrelated at the level of overall sequence similarity may perform the same function in another species (see p. 152). A third complication is that a single protein in one species may perform functions that are performed by more than one protein in other species.

Data Sources, Methods, and References

The figure was generated as described in chapter 1. The sequences used in the searches for the figure were GI:4506567 (for RNMT), GI:38150007 (for SNRPB), GI:71143123 (for C6orf151), GI:61743957 (for APOBEC1), and GI:4501919 (for ADARB1). SNRPB does not align simply with its *Drosophila* counterpart.

See also:

Conticello S.G. et al. 2005. Evolution of the AID/APOBEC family of polynucleotide (deoxy)cytidine deaminases. *Mol. Biol. Evol.* **22:** 367–377.

Gerber A. et al. 1998. Tad1p, a yeast tRNA-specific adenosine deaminase, is related to the mammalian pre-mRNA editing enzymes ADAR1 and ADAR2. *EMBO J.* **17:** 4780–4789.

Will C.L. et al. 2004. The human 18S U11/U12 snRNP contains a set of novel proteins not found in the U2-dependent spliceosome. *RNA* **10:** 929–941.

How Does Protein Sequence Conservation Differ Among the Histone Types?

Histone types vary considerably with regard to sequence conservation across the eukaryotes. In the figure below, one protein sequence from each of the major histone classes encoded in the large cluster on human chromosome 6 (see p. 101) was used to search for the best matches in diverse eukaryotes.

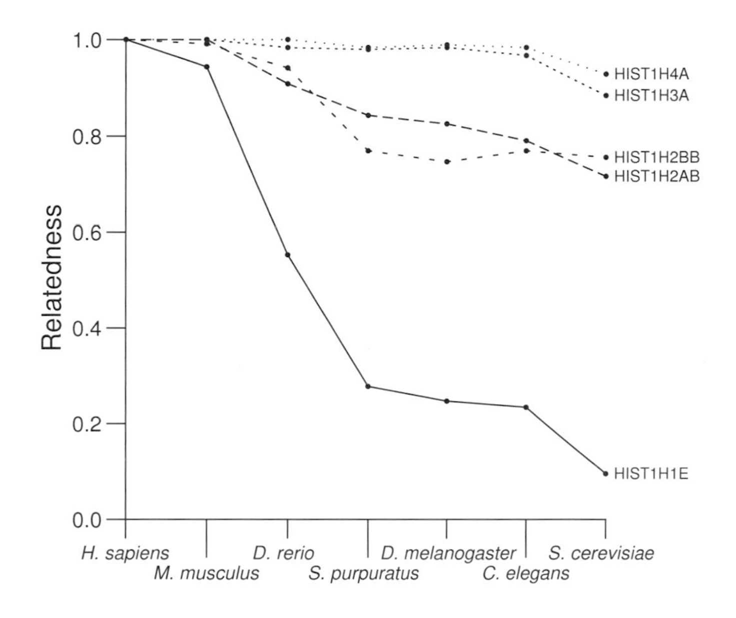

As can be seen in the figure, histones H3 and H4 show extraordinary conservation, with only limited divergence in *S. cerevisiae*. This is comparable to the level of conservation in cytoplasmic actin. H2A and H2B are relatively well-conserved proteins. In contrast, H1, with its low-complexity regions, varies considerably within the vertebrates, and its yeast counterpart does not produce simple alignments with the human protein.

Data Sources, Methods, and References

The following protein sequences from RefSeq were used: GI:4885379 (HIST1H1E), GI:19557656 (HIST1H2AB), GI:10800140 (HIST1H2BB), GI:4504281 (HIST1H3A), and GI:4504301 (HIST1H4A). The graph was produced as described in chapter 1.

How Conserved Are the Branches of the Ras Superfamily?

There are five main branches of the Ras superfamily. One human protein from each branch was used to search for proteins encoded by other genomes. The figure below shows the results.

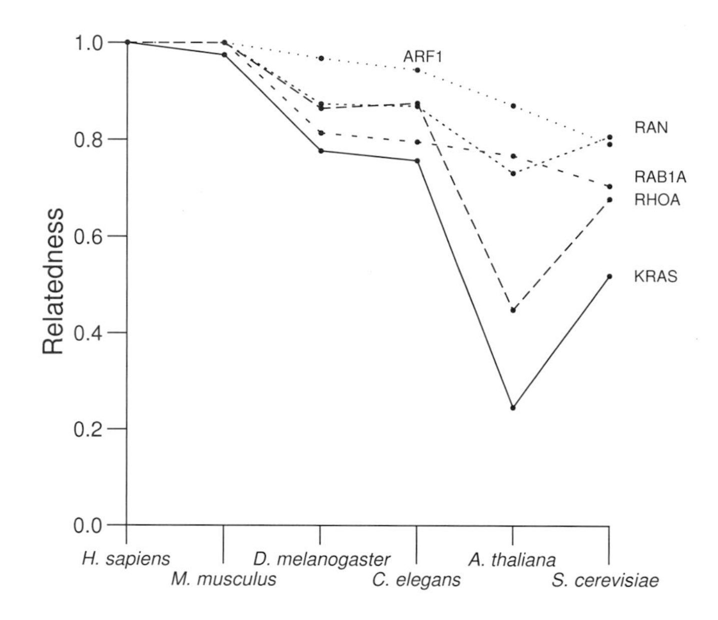

All queries readily identified closely related proteins from diverse eukaryotes. Except for two cases involving *A. thaliana* (RHOA and KRAS), all show sequence divergence that is reasonably well-correlated with evolutionary distance. All five subfamilies are present in *S. cerevisiae*, with the Ras types (represented here by KRAS) being the least conserved. The ADP ribosylation factor ARF1 is highly conserved in the eukaryotes.

In *A. thaliana*, representatives of the Ras branch are not present. The match in the figure (for KRAS) is to a protein in another branch of the *A. thaliana* Ras superfamily. *A. thaliana* does have members of the Rho branch (represented here by RHOA), but they are closer to the human Rac types (which are also in the Rho branch) than to the Rho types.

Data Sources, Methods, and References

The figure was generated using the method described in chapter 1. Each of the proteins named in the figure was used to search complete protein sets from each of the species. The query sequences were GI:5453555 (RAN), GI:4758988 (RAB1A), GI:10835049 (RHOA), GI:15718761 (KRAS), and GI:4502201 (ARF1). The best match is reported, even if it is not an orthologous protein.

See also:

Vermoud V. et al. 2003. Analysis of the small GTPase gene superfamily of *Arabidopsis*. *Plant Physiol.* **131:** 1191–1208.

How Do the Sizes of Gene Families for Structural Proteins Vary Among the Eukaryotes?

Many of the major human structural proteins are encoded by large gene families. Such large families are not uniformly present in multicellular species, and these families often consist of only one or two genes in yeast. The table below presents the number of genes in some of the most widely studied gene families based on their RefSeq entries.

	No. of Genes in Family		
	Actin	Myosin (heavy)	α-tubulin
Human	6	~37	9
D. melanogaster	7	13	5
C. elegans	5	17	9
S. cerevisiae	1	5	2

Even for such well-known structural proteins, assembling accurate gene counts is difficult (see the notes and exceptions at the end of this section). These genes are parts of larger families with additional members. For example, the genes for centractins, which are not included in the counts above, are quite similar to the principal actin types.

Data Sources, Methods, and References

The human gene counts were from NCBI RefSeq protein entries except for the addition of MYH16 (see GI:10438291). The related genes in the other species were from BLASTP searches of RefSeq entries for their protein products. Annotated pseudogenes and predicted fragments have been excluded from the total gene counts.

Actin genes were defined as being more similar to the six main human types than to centractin. This resulted in the inclusion of the more diverged *CG5409* gene from *Drosophila* and the *T25C8.2* gene from *C. elegans*.

Myosins include all the various heavy chain types and the related unconventional myosins, including the more distant *Drosophila CG10595* gene. Excluded is the somewhat smaller but related *CG5939* gene in *Drosophila*.

α-tubulins are generally straightforward to identify because they are considerably more related to each other than to the β-tubulins. The more diverged *Drosophila* tubulin gene *CG7794* was included in the table.

How Large Are Gene Families for Developmental Signaling in Animals?

Hedgehog, Notch, and Wnt are three of the main signaling pathways in animal development. The pathways are named for intensively studied genes in *Drosophila* that encode the main components of the pathways. The table below presents the number of genes in these families in human, *D. melanogaster*, and *C. elegans*. Protein-level searches were used to identify related genes. The counts are from RefSeq entries.

	No. of Genes in Family		
	Hedgehog	Notch	Wnt
Human	3	4	19
D. melanogaster	1	1	7
C. elegans	~10	2	5

Although human families for genes of this type are typically larger than those in the other two species, the hedgehog genes are a notable exception. In most animals, hedgehog proteins are generally quite similar to each other and are unlike other proteins. However, in *C. elegans*, the hegdehog-related proteins are considerably less similar to the corresponding human proteins, vary considerably in size, and fall into several subtypes. Additional searches against all of the *C. elegans* RefSeq entries using the ten *C. elegans* proteins counted in the table (found directly via searches with their human relatives as queries) identify even more family members.

The Notch proteins contain domains found in numerous other proteins. To identify the counterparts in the other species, a fragment of NOTCH1 was used as a query in the searches, as detailed below.

The Wnt family members were readily distinguished from other proteins in BLASTP searches.

Data Sources, Methods, and References

Searches used NCBI BLASTP. Pseudogenes and predicted gene fragments were excluded from the totals.

A fragment of human NOTCH1 (GI:27894368; amino acids 1401 to 1900) was used to search for related proteins in the other species. This query readily distinguished the Notch proteins from more distant relatives, such as the jagged family. The two Notch-related genes identified in *C. elegans* were *lin-12* and *glp-1*.

For more extensive tables of this type from a draft genome sequence, see:

Venter J.C. et al. 2001. The sequence of the human genome. *Science* **291:** 1304–1351.

Which Species Have Genes Related to Those with Functions in Human Host Defense?

Genes related to human host defense functions can be found in surprisingly diverse species. For example, toll-related receptors are widely found in the eukaryotes. To search for such genes in other species, protein-level searches can be used as surrogates. The figure below shows the results of searches using proteins from four pathways: RAG1 (antigen receptor recombination), TLR3 (a toll-type receptor), CFB (factor B in the alternate complement pathway), and MEFV (pyrin; the Mediterranean fever locus).

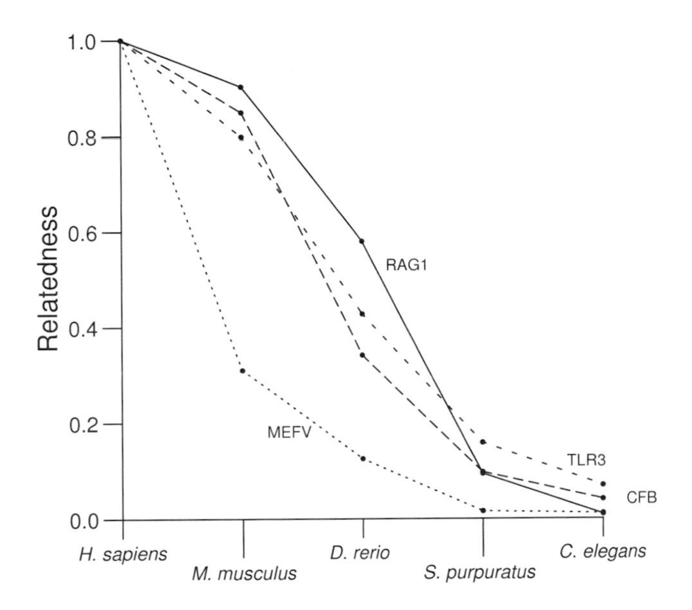

Proteins related to RAG1, CFB, and TLR3 are encoded in more distant species such as the sea urchin (*S. purpuratus*). In contrast, the MEFV protein is quite different, even in mice. Results such as those in the figure above are not always straightforward to interpret: The sea urchin CFB-related protein is also related to human C2. The weak matches for TLR3 and CFB in *C. elegans* are actually to proteins that, when used in a reverse search, are more closely related to other human proteins (not human TLR3 or CFB). The *MEFV*-related genes in zebrafish encode other tripartite motif-containing proteins.

Data Sources, Methods, and References

The figure was produced by the methods described in chapter 1. The sequences used as queries were GI:4557841 (for RAG1), GI:4507531 (for TLR3), GI:67782358 (for CFB), and GI:4557743 (for MEFV).

See also:

Rast J.P. et al. 2006. Genomic insights into the immune system of the sea urchin. *Science* **314:** 952–956.

Which Human Proteins Are Related to Retroviral Proteins?

A large number of human genes encode proteins that are related to products of retroviral oncogenes. Many of these proteins are listed in the table below.

Human Protein	Function	Some Related Human Proteins
ABL1	tyrosine kinase	ABL2
AKT1	protein kinase	AKT2, AKT3
CBL	ring finger protein	CBLB, CBLC
CRK	SH2 and SH3 domains	CRKL
CSF1R	colony-stimulating factor 1 receptor	
EGFR	epidermal growth factor receptor	ERBB family
ETS1	transcription factor	ETS2
FES	tyrosine kinase	FER
FGR	tyrosine kinase	
FOS	transcription factor	FOSB, FOSL1, FOSL2
FOXG1B	transcription factor	
HRAS	Ras family GTPase	NRAS, KRAS (also listed below)
JUN	transcription factor	JUND, JUNB
KIT	mast cell factor receptor	
KRAS	Ras family GTPase	NRAS, HRAS (also listed above)
MAF	transcription factor	MAFB, MAFA
MOS	protein kinase	
MPL	thrombopoietin receptor	
MYB	transcription factor	MYBL1, MYBL2
MYC	transcription factor	MYCN, MYCL1
PDGFB	platelet-derived growth factor β	PDGFA
RAF1	protein kinase	BRAF, ARAF
REL	transcription factor	RELA, RELB, NFKB1, NFBK2
ROS1	receptor tyrosine kinase family	
SKI		SKIL
SRC	tyrosine kinase	
THRA	thyroid hormone receptor α	THRB
YES	tyrosine kinase	

Retroviral protein sequences that are related to the human proteins listed in the left column are detailed in the notes at the end of this section. Reverse searches with these viral protein sequences frequently identify other related human proteins. Some notable examples of these related proteins are included in the right column. In certain cases, such as the tyrosine kinases, the families are quite large.

The results of the searches with the viral protein sequences can reveal relationships within some of the larger families. For example, for the Ras family GTPases, the viral counterparts of HRAS and KRAS clearly distinguish between those human proteins and the third family member, NRAS.

Although portions of these viral gene products are often highly conserved relative to human proteins, the overall size and structure of any given viral protein may be quite different than its human relative.

Data Sources, Methods, and References

The viral sequences were identified using the precalculated BLASTP results (BLink) at NCBI for the RefSeq entries for the products of the human genes listed in the left column of the table. The entries in the right column were obtained by performing BLASTP searches of human RefSeq entries with these viral sequences.

The NCBI identifiers for the viral protein sequences (and the symbols for the corresponding human proteins listed in the left column of the table) are GI:5912560 (ABL1), GI:210068 (AKT1), GI:323270 (CBL), GI:61493 (CRK), GI:125363 (CSF1R), GI:6016434 (EGFR), GI:223807 (ETS1), GI:209689 (FES), GI:61543 (FGR), GI:120472 (FOS), GI:11178689 (FOXG1B), GI:131872 (HRAS), GI:209724 (JUN), GI:224986 (KIT), GI:131877 (KRAS), GI:209628 (MAF), GI:332209 (MOS), GI:332298 (MPL), GI:127590 (MYB), GI:323923 (MYC), GI:61777 (PDGFB), GI:4388778 (RAF1), GI:136184 (REL), GI:9627733 (ROS1), GI:134516 (SKI), GI:125706 (SRC), GI:209669 (THRA), and GI:223452 (YES).

HUMAN GENE INDEX

This index covers human protein-coding genes mentioned by name in the text or shown in the figures. Human genes for ncRNAs and pseudogenes are not included. Note that there are often many synonyms for human gene names in the literature. Alternate names indicated in the notes and gene names in reference titles cited in the text have also been excluded. The gene names are sorted character by character (e.g., *MUC19* precedes *MUC2*).

A

B

C